Cognitive Science and Technology

Series editor

David M.W. Powers, Adelaide, Australia

T0189644

More information about this series at http://www.springer.com/series/11554

Jochen Huber · Roy Shilkrot
Pattie Maes · Suranga Nanayakkara
Editors

Assistive Augmentation

 Springer

Editors
Jochen Huber
Synaptics
Zug
Switzerland

Roy Shilkrot
Department of Computer Science
Stony Brook University
New York, NY
USA

Pattie Maes
Massachusetts Institute of Technology
   (MIT)
Cambridge, MA
USA

Suranga Nanayakkara
Augmented Human Lab
Singapore University of Technology and
   Design
Singapore
Singapore

ISSN 2195-3988   ISSN 2195-3996   (electronic)
Cognitive Science and Technology
ISBN 978-981-13-4872-3   ISBN 978-981-10-6404-3   (eBook)
https://doi.org/10.1007/978-981-10-6404-3

Printed on acid-free paper

This Springer imprint is published by Springer Nature
The registered company is Springer Nature Singapore Pte Ltd.
The registered company address is: 152 Beach Road, #21-01/04 Gateway East, Singapore 189721,
Singapore

Contents

Introduction

Jochen Huber, Roy Shilkrot, Pattie Maes and Suranga Nanayakkara

> *My years of experience at Dr. Reijntjes School for the Deaf, Sri Lanka made me realize that sensory impairment has nothing to do with intellectual ability. For instance, the deaf children at this school, were able to communicate over much longer distances with sign language and make beautiful computer graphics. In fact, they had such a developed special skill that I felt like the odd man.*
>
> Suranga Nanayakkara

Our senses are the dominant channel for perceiving the world around us. With impairments and lack thereof, people find themselves at the edge of sensorial capability. Some excel and use their impairment as a gift. Prominent examples are Evelyn Glennie, a percussionist with hearing impairment, and Ben Underwood, whose eyes were removed when he was 5 years, taught himself echolocation. Some seek assistive or enhancing devices which enable a "disabled" user to carry out a task or even turn the user into a "superhuman" with capabilities well beyond the ordinary. The overarching topic of this volume is centered on the design and development of assistive technology, user interfaces and interactions that seamlessly integrate with a user's mind, body and behavior in this very way–providing enhanced physical, sensorial and cognitive capabilities. We call this "Assistive Augmentation".

J. Huber
Synaptics, Zug, Switzerland
e-mail: jochen.huber@acm.org

R. Shilkrot
Stony Brook University, New York, USA
e-mail: roys@cs.stonybrook.edu

P. Maes
MIT, Cambridge, USA
e-mail: pattie@media.mit.edu

S. Nanayakkara (✉)
Singapore University of Technology and Design, Changi, Singapore
e-mail: suranga@ahlab.org

© Springer Nature Singapore Pte Ltd. 2018
J. Huber et al. (eds.), *Assistive Augmentation*, Cognitive Science
and Technology, https://doi.org/10.1007/978-981-10-6404-3_1

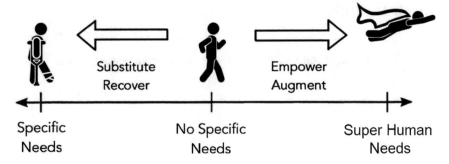

Fig. 1 Assistive augmentation continuum, a guideline for developing cross-domain assistive technology

Assistive Augmentation finds its application in a variety of contexts, for example scaffolding people, when they feel their innate senses are inadequate or, to support development of desired skillsets. We wish to put sensorial capability on a continuum of usability for certain technology, rather than treat one or the other extreme as the focus (cf. Fig. 1).

We therefore follow the design rationale of [1], stating technology should be socially acceptable, work coherently for disabled and non-disabled alike, and support independent and portable interaction. The latter requirement challenges both user interface and interaction design in particular, as Jones and Marsden point out: *"the test comes when it [the device] is deployed in the complex, messy world of real situations [...] when it has to be used in the wild, as it were, in tandem with the world around it, the usability can break down quickly"* (cf. [2] p. 51). In the following, we depict challenges for the field of assistive augmentation and outline the objectives and the structure of this volume.

1 Challenges for Assistive Augmentation

The design, implementation and deployment of assistive augmentation technology faces a variety of challenges due to its cross-disciplinary nature. Emerging technologies for human augmentation continuously change how we perceive and interact with our surroundings, as well as ourselves. They strive to augment our sensory abilities for increased well-being, e.g. by stimulating our motor system [3] or even the gustatory perception [4]. At the same time, assistive technologies emerge that promise e.g. to scaffold sensory disabilities, e.g. to improve reading capabilities of the blind through technologies such as BrailleTouch [5] or Ubi-Braille [6].

Existing ethnographic research sheds light on stigma and misperceptions people face when using assistive technology in social situations [7], such as being publicly marked as a disabled person, or that technology can effectively eliminate disability.

Augmenting technology also harbors ethical implications, as it breaks the conception of an even playing field for all, once certain people start augmenting their natural-born senses with technology [3].

This triggers more higher level questions such as what is a good assistive augmentation, what is its quality and when can it be considered successful? Also, can we build on well-established research and evaluation methods that are effective in the assistive technologies and accessibility communities? How can technology be designed to discourage stigma, self-consciousness or social asymmetry in its users?

2 Objectives and Structure of This Volume

Research on Assistive Augmentation is spread across many different communities, depending on the targeted part of the usability continuum sketched above. Accessibility and assistive technologies are topics pertinent to academic conferences such as ACM ASSETS, and also widely disseminated at the CHI main conference. On the other hand, human augmentation is a primary topic at conferences such as Augmented Human or the International Symposium on Wearable Computing. The idea of Assistive Augmentation is also emerging in the field of sports, with the vision of reinventing sports that anyone can enjoy regardless of the ability.

Addressing this disparity, the Assistive Augmentation community initially met at an interdisciplinary workshop at the 2014 International Conference on Computer Human Interaction (CHI 2014) in Toronto, Canada [8]. The community is comprised of researchers and practitioners who work at the junction of human-computer interaction, assistive technology and human augmentation. This edited volume is the first tangible outcome of this very workshop.

The goal of this edited volume is to illustrate core areas of Assistive Augmentation and to stimulate discussion around challenges within those. As a first step, this volume explores (i) Sensory Enhancement & Substitution and (ii) Design for Assistive Augmentation. Implemented research within these areas directly caters to the continuum depicted in Fig. 1, investigating specifically:

- Development of novel technologies or extension of available technologies to synthesize desired augmentations or enhancements
- Understanding perceptual, sensorial, cognitive, and behavioral capabilities of users and discovering sensory substitution strategies
- Exploration of the interaction design space with proof-of-concept prototypes

The remainder of this volume contains comprehensive reports on case studies that focus on either of the core areas. The studies serve as lighthouse projects, each of them contributing to a sub-issue and challenge of Assistive Augmentation. Clustered into respective areas, these also map to the structure of this volume:

Part 1: Sensory Enhancement & Substitution

Chapter 3: Contributes sound-to-vibrotactile sensory substitution systems for deaf people in the application areas of music listening and music making.
Chapter 4: Describes an Augmented Reality system that focuses assisting human workers in flexible production environments.
Chapter 5: Reports on narrative text augmentations with sound effects to provide embodied experiences to readers.

Part 2: Design for Assistive Augmentation

Chapter 7: Discusses how technologies and domestic environments need to be designed to support ageing in place.
Chapter 8: Contributes a design process of a responsive sensory environment that augments social communication between autistic children and their parents.
Chapter 9: Illustrates the extensive design process of a finger-worn device equipped with a camera that assists users with visual impairments in accessing printed text.

Both parts of this volume are prefaced by introductions to the respective area of research. The volume concludes with a summary and an outlook upon the future of Assistive Augmentation.

References

1. Rekimoto J, Nagao K (1995) The world through the computer: Computer augmented interaction with real world environments. In: Proceedings of the 8th annual ACM symposium on User interface and software technology, pp 29–36
2. Jones M, Marsden G (2006) Mobile interaction design. Wiley
3. Liao T (2012) A framework for debating augmented futures: classifying the visions, promises and ideographs advanced about augmented reality. In: 2012 IEEE international symposium on mixed and augmented reality (ISMAR-AMH), pp 3–12
4. Nakamura H, Miyashita H (2011) Augmented gustation using electricity. In: Proceedings of the 2nd augmented human international conference, New York, NY, USA, pp 34:1–34:2
5. Romero M, Frey B, Southern C, Abowd GD (2011) BrailleTouch: designing a mobile eyes-free soft keyboard. In: Proceedings of the 13th international conference on human computer interaction with mobile devices and services, New York, NY, USA, pp 707–709
6. Nicolau H, Guerreiro J, Guerreiro T, Carriço L (2013) UbiBraille: designing and evaluating a vibrotactile braille-reading device. In: Proceedings of the 15th international ACM SIGACCESS conference on computers and accessibility, New York, NY, USA, pp 23:1–23:8
7. Shinohara K, Wobbrock JO (2001) In the shadow of misperception: assistive technology use and social interactions. In: Proceedings of the SIGCHI conference on human factors in computing systems, New York, NY, USA, pp 705–714
8. Huber J et al (2014) Workshop on assistive augmentation. In: CHI '14 extended abstracts on human factors in computing systems, New York, NY, USA, pp 103–106

Part I
Sensory Enhancement and Substitution

Augmented Sensors

Suranga Nanayakkara, Jochen Huber and Priyashri Sridhar

Key elements of the emerging field of Assistive Augmentation are both the sub-stitution and enhancement of senses–means towards *"augmented sensors"*. We use the very term "augmented sensors" to introduce the following subsections of this part of the volume that focuses on enhancing a particular sensory channel, remapping information from one sensory modality to another and creating new sensing modalities (cf. Fig. 1). We do so by describing our vision of such tech-nology developed at the Augmented Human Lab,[1] sketching out research thrusts and enablers, highlighting application domains and speculating about the future of augmented sensors.

1 Research Thrusts and Enablers

We exemplify the challenges for Augmented Sensors through pertinent research conducted at the Augmented Human Lab (Singapore University of Technology and Design). This work is along three highly interdisciplinary research thrusts (Fig. 2): (1) Novel User Input and Interaction Techniques, (2) Sensory Substitution and Fusion Technology, (3) Cognitive Augmentation.

---

[1]http://www.ahlab.org.

S. Nanayakkara (✉) · P. Sridhar
Singapore University of Technology and Design, Changi, Singapore
e-mail: suranga@ahlab.org

J. Huber
Synaptics, Zug, Switzerland

© Springer Nature Singapore Pte Ltd. 2018
J. Huber et al. (eds.), *Assistive Augmentation*, Cognitive Science
and Technology, https://doi.org/10.1007/978-981-10-6404-3_2

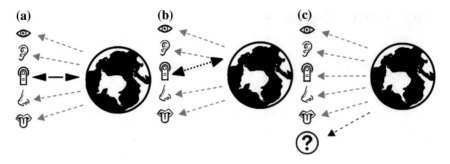

Fig. 1 Modes of augmented sensors: **a** enhancing a particular sensory channel; **b** remapping information from one sensory modality to another; **c** creating new sensing modalities

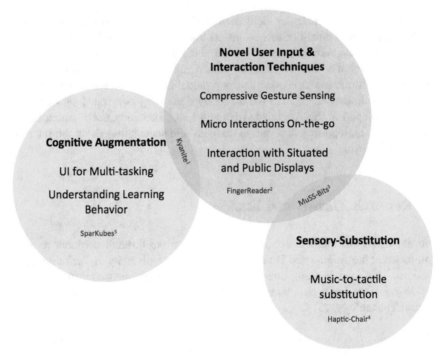

Fig. 2 Research thrusts of augmented sensors in the Augmented Human Lab (http://www.ahlab. org/projects). Example projects are listed in blue color (*1* http://www.ahlab.org/project/kyanite, *2* http://www.ahlab.org/project/fingerreader, *3* http://www.ahlab.org/project/muss-bits, *4* http:// www.ahlab.org/project/hapticchair, *5* http://www.ahlab.org/project/sparkubes)

1.1 Novel User Input and Interaction Techniques

Current computer systems lack the contextual knowledge to offer relevant information at the right place and time. They are more like a tool, a hammer for instance —when you need to get some work done, you use the tool and provide explicit instructions to it. On the contrary, what if the tool could potentially guide you on what to do? What if your smartphone is able to inform you that you owe your friend $5 when you meet him? To tackle this, we need to research on developing new ways to interact with computers (i.e. user inputs and interactions). For example, with the advancements of affective computing, deep learning neural networks, and power of GPU, researchers have developed a tool that is capable of understanding the user in a more holistic way.[2] Perhaps researchers from various fields including interaction design, machine learning, and ubiquitous computing would have to leverage on these to move to a paradigm outside of the 'computer box'. In fact, introduction of virtual reality devices such as the Oculus Rift and the Samsung GearVR requires interaction methods go beyond the traditional touch/button based interfaces. In addition, augmented reality interfaces such as Microsoft Hololens bring us outside of our world into a detached reality. Given the physical space and energy constraints, we have to look beyond computer vision based gesture recognition techniques. New technologies (such as zSense [1]) are needed to increase the input expressivity of such resource restricted devices.

1.2 Sensory Substitution and Fusion Technology

Drawing inspiration from novel ways of interacting with tools, one can imagine how limitless our capabilities would be if we could use these novel interactions to augment our sensory abilities. What if we were able to temporarily extend our field of view towards 360°, allowing us to see things happening around us and to anticipate a dangerous situation happening behind us? What if a person with deafness could perceive previously inaccessible auditory information through vibro-tactile feedback? We explored the former question in SpiderVision [2], a head mounted display that enhances the human field of view for augmented awareness and the latter in works such as Haptic Chair [3]. Such approaches, we believe, will empower people to use the available communication bandwidth between our senses effectively or even increase it.

Key to this approach is (i) sensory augmentation technology that makes individual senses more accurate and effective, (ii) sensory substitution technology that

---

[2]https://www.soulmachines.com/.

remaps sensory information (Haptic Chair [3] and Music Sensory Substitution (MuSS) Bits [4]), and (iii) fusion technology that has the power to formulate new sensory modalities that expand upon vision, hearing, touch, smell and taste (Taste+ [5]).

1.3   Cognitive Augmentation

While it is fascinating to have new ways of interacting with the environment or integrating our sensory modalities to enhance our performance, they may impose some effort on our part. Further, we live in an era that requires us to constantly multi-task between a variety of activities often leaving us overwhelmed due to the overload of information. Even as you read this paragraph, you are using attention and memory. While information and tasks can be limitless, there is a limit on the cognitive processes that humans possess, particularly, attention and memory. The amount of cognitive resources depends on the difficulty of the task as well as the number of tasks that are performed concurrently. The more complex the task or greater the number of tasks/items to be remembered, the higher will be the cognitive load [6]. Through Cognitive Augmentation, we seek to understand a user's cognitive state and develop technologies that help users make more informed decisions with less cognitive effort. The knowledge gap we need to fill is the holistic understanding of the possibilities of merging different modalities afforded by the technological advancements. This is typically approached by designing and systematically refining a prototype system with a series of end-user experiments. We use a triangulated framework of objective and subjective approaches to study the user's cognitive state as they perform a variety of tasks in controlled and natural environments. Drawing from research in psychology, neuroscience and information technologies, this emerging field has several implications in defence services, rehabilitation and education.

1.4   Enablers: Technology and Design Innovation

Design for Acceptance: The success of any technology is determined by the ease of acceptance and use by the user community. This is more crucial when designing assistive technologies in any form, given the intention behind their creation. The cultural and experiential gap between researchers and end users can be especially large when developing such assistive technologies. Such a gap can lead to a situation where developers make products solely based on their own interpretation of the needs, a solution that can be ineffective and patronizing. However, adopting a "User Sensitive Inclusive Design" process [7], which includes identification of specific techniques for eliciting information from the target user group and strategies for involving them in user experience studies overcomes this gap to a large extent. Specially tailored focus groups discussions, semi-structured in-depth interviews, and in-home observations designed to study the usability and user experiences of the applications produce iterative results that effectively contribute the hardware and software development and vice versa.

Customized Hardware and Software: Off-the-shelf hardware covers a wide range of modalities. However, they are being developed with particular computing paradigms in mind e.g., for applications in robotics or consumer electronics. Both software and hardware development that pertain to the research thrusts described above have unique requirements. As such, it is critical to develop customized hardware and software to prototype application scenarios of assistive devices (for example, our prior work FingerReader [8], Bward [9]). Such assistive devices have been designed using custom made printed circuit boards (PCBs), emerging sensing technologies [10] and communication mediums (e.g. Low Power WAN) and additive manufacturing technology resulting in hardware prototypes that can be adapted to dynamic requirements.

2   Application Domains

In light of the research thrusts and enablers discussed above, we identify some potential implementation scenarios and outline some of the practical applications that the Augmented Human Lab has been working on. While some research thrusts find a direct implementation in some of the cases listed below, some of our work lies at the intersection of these research thrusts. A common thread underlying these applications are the enablers—customized solutions borne out of a user-centered design process. These projects illustrate the potential that augmentation holds in enabling changes across diverse communities and capabilities.

2.1   Independent Living for the Ageing Population

Sustaining the capabilities, independence and resourcefulness of older adults, and helping them to age gracefully, is a key challenge we are facing now. Traditionally, technologies developed to improve the lives of the elderly have mainly focused on physiological needs and safety concerns. We believe that the opportunities for technology do not just lie in memory, cognition and communication but also in sustaining the identity, self-reliance and self-worth of an individual. As such, we aim to design, develop and implement technologies that empower older adults and help them sustain their resourcefulness and independence that can make a significant difference.

For instance, we developed StickEar [11] (Fig. 3), a wireless, re-deployable and reconfigurable sound-based sensor to empower older adults to create a local

Fig. 3 StickEar [11] prototype

wireless sensor network at home. It could, for example, help an elderly person suffering from degenerative hearing loss to know if someone is knocking at the door or if water is boiling in a whistling kettle. This technology can be extended to other types of sensors such as gas, water, temperature, etc. StickEar can also be used as an output device, allowing a user to trigger a sound output on StickEar from their mobile device. The elderly can use this to locate objects that they have misplaced by simply speaking into their mobile device and triggering an alarm sound on the StickEar that is attached to that object.

In another project, WatchMe [12], we capitalized on the concept of remote sensing to understand the living behaviour of the elderly and use the information to alert family in cases of emergency. The WatchMe system (Fig. 4) is implemented on a regular smartwatch with the focus on making ambient monitoring intuitive and seamless. For instance, a Caretaker's WatchMe can be paired with the WatchMe of the person who needs support, using a simple tap gesture. In addition to the ease of pairing and switching among different caretakers, the wristwatch interface allows users to simply glance at their smartwatch to get a sense of the state of the remote user. We believe, these types of seamless interactions would create a healthy link between them and their loved ones who might have busy schedules.

2.2 Assistive Technology for the People with Visual Impairments

It is estimated that about 285 million of the world's population have some form of visual impairment. While the severity of the condition for people with visual impairments varies from individual to individual, they still lack in independence

Fig. 4 WatchMe [12] prototype

and the proper technology to aid in everyday tasks. The major hurdles that persons with visual impairment face are (i) affordability, (ii) usability and (iii) social acceptance. Related technologies available on the market come with a price tag in the order of thousands of dollars (e.g. OrCam at $2,500), require heavy instrumentation, involve a steep learning curve and are usually bulky. The latter also brands its users as "special needs" persons. We observe that finger-worn interfaces remain an unexplored space for assistive user interfaces, despite the fact that our fingers and hands are naturally used for referencing and interacting with the environment. As such, we focused on developing a finger-worn interface to support a blind person in everyday tasks.

As a starting point, we designed and developed FingerReader [8] (Fig. 5) to assist blind users with reading printed text on the go. We introduce a novel computer vision algorithm for local-sequential text scanning that enables reading single lines, blocks of text or skimming the text with complementary, multimodal feedback. This system is implemented in a small finger-worn form factor that enables a more manageable eyes-free operation with trivial setup. The perpetual, broad media coverage of our line of finger-worn devices underlines the significance of the problem at hand, concerning an important community within our society. We plan to develop proof of concept assistive technologies for people with sensory disabilities to be more independent in their way finding and play a more active role in social relationships.

2.3 Assistive Technology for the Deaf Community

Our work with communities having sensory disabilities extends beyond those with visual impairments. It is estimated that over 5% of the world have some form of disabling hearing loss across age groups thereby affecting their ability to perceive

Fig. 5 FingerReader [8] prototype

different forms of speech and music in the environment. We explored the possibility of translating music, an auditory signal into vibro-tactile feedback through Haptic Chair [3]. Haptic Chair (Fig. 6) is a sensory substitution interface that translates music into vibro-tactile feedback, providing rich musical experiences to deaf users via 'full body haptic stimulation'.

In order to better understand how the system works in a more natural environment, we deployed this system in a residential deaf school to be used on a daily basis. It was encouraging to get positive feedback on how this form of sensory substitution enabled even profoundly deaf users to "hear" a song.

Fig. 6 HapticChair [3] prototype

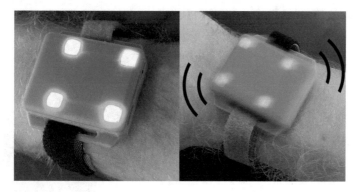

Fig. 7 MussBits [4] prototype

Inspired by this, we developed Music Sensory Substitution (MuSS) Bits [4] (Fig. 7), small wearable plug-and-play sensor-display pairs that capture real-world sounds, extract the rhythm information and convert them into visual and vibrotactile output. We deployed a working prototype of MuSS Bits in the same school, focusing on conveying rhythm information to deaf performers. Our studies demonstrate its effect in improving rhythm recreation for deaf children.

2.4 Sensing and just-in-Time Information for Smart Health

Health care professionals report on numerous shortcomings of existing bedside care systems: (i) most commercial systems provide only reactive support, (for example, alert the clinical staff when the patient has already fallen); in addition these alarms are too disturbing—especially at night; (ii) most systems produce a high rate of false alarms and therefore lack in both effectiveness and efficiency; (iii) although false positives are preferable over false negatives, they increase alarm fatigue and the average reaction time of the clinical staff significantly. Many design opportunities exist that go beyond reactive support. As such, with context-aware wearable and tangible interfaces, we aim to explore new ways of managing bedside care.

As a first step, we have explored design opportunities together with stakeholders (Doctors, clinical staff, patients) at Changi General Hospital in Singapore adopting a human-centered design process. We designed and developed a robust and reliable in situ early blood leakage detection device, BWard [9] (Fig. 8), which is tailored for clinical needs and environment in CGH hospital. This novel system consisted of a reliable detection system along with a programmable audible and visual alarm system and was integrated seamlessly with the ward/nurse call monitoring systems. This system could eliminate the requirement of medical staff having to manually observe a wound site after dialysis catheters are removed.

Fig. 8 BWard [9] prototype

2.5 Personalized and Continuous Rehabilitation

Rehabilitation training typically involves extensive repetitive range-of-motion and coordination exercises. This requires a substantial effort from a therapist to supervise and assess the progress of a patient. However, in most cases the rehabilitation process cannot be performed with sufficient intensity due to limited human and financial resources [6]. Further, existing systems are typically bulky, complicated and has ergonomically poor in design (e.g. Sun SPOT sensor node [13], wearable sensors interconnected by wires [14], etc.). To overcome the limitations, we augment the current rehabilitation processes with responsive objects and serious gaming to increase motivation and provide personalised care. This includes physical/virtual rehabilitation game design, non-intrusive sensing device design, sensing system design and data analytics.

As one instantiation, our team developed a proof-of-concept prototype, SHRUG [15] (Fig. 9), in consultation with the medical professionals dealing with stroke rehabilitation at St Andrew's Community Hospital, Singapore. This has two elements: (1) a main rehabilitation device, based on the hospital apparatus enhanced with a sensor and a feedback system and (2) a pole interface designed to interact with the main device. The pole interface provides a sense of ownership and enhances the gaming element, as the pole interface will display the users' score and 'belongs' to the user as a personal device across potentially multiple rehabilitation devices. A complementary information dashboard was developed to provide therapists, access to performance data to enable personalized care. With this system, we expect a real-world demonstration of providing interactive and gamified feedback to engage the patients and empower the therapists to provide personalized care.

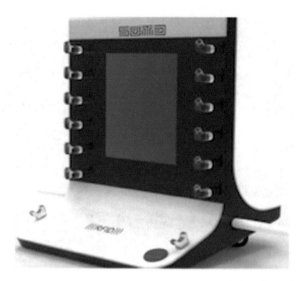

Fig. 9 SHRUG [15] prototype

2.6 Interfaces to Support Learning

Learning does not occur in a vacuum. Any learning process typically consists of learners and learning tools or objects in the environment that the learner interacts with. In many cases, an adult/teacher who guides and enables the learning process may also be present. The learner himself is equipped with two of the states that affect learning behavior and outcome to a great extent [16]: (1) cognitive state and (2) affective/emotional state. Cognitive state includes executive functioning such as working memory, inhibition and flexibility. But this cognitive state is affected by emotional states that inturn affects learning. By understanding these underlying states during learning, we can design interfaces that best enable learning across age groups. Research has shown that when physical objects become more interactive, when physical components are interactive, they result in more engagements and become more playful. As a preliminary work, we explored play behaviour in children to understand how normal blocks can be made interactive and the subsequent influence of such an addition on play dynamics. Through free play sessions, we observed the patterns that children formed using normal blocks versus Spar-Kubes [17]. SparKubes (Fig. 10) are a set of stand-alone tangible objects that use the flow of light as the principle of operation. They are corded with simple behaviors and they do not require any special instrumentation or setup. Our observations revealed that children not only tend to spend more time exploring the interactive features but also formed a greater variety of patterns using SparKubes [18] as compared to the normal blocks. Our findings also revealed that making normal objects interactive has the potential to increase the play value of an object, thereby making the interaction more engaging.

Fig. 10 Sparkubes [17] prototype

2.7 Interactive Media for Community Engagement

In the urban public arena, media platforms can serve as a space of creative and artistic engagement between people, exploring and building a sense of belonging and community. For example, SonicSG [22] has a kind of "double ontology" [23] with both a visual/sonic/interactive aesthetic dimension situated in an urban recreational river walkway. Visitors were invited to participate in the work by pointing their mobile device browsers to sonic.sg. There they entered the postal code of their Singapore neighbourhood. After writing a birthday wish and submitting it, a pebble-drop ripple of light emanated from their neighbourhood location in the floating light display. This effect provided immediate feedback and public evidence of their participation, creating a connection between the audience and the work. Another "layer of connectedness," in the audience was then established by turning their mobile phones into a distributed array of "sonified personal pixels." Each phone slowly pulsed a color and tone unique to their neighbourhood at a rate that was a function of the number of other participants from the same neighbourhood. As participants moved around and explored the installation, a light and sound texture was created among the audience, reflecting both the dynamic diversity of neighbourhoods and the unified tapestry they collectively comprise as a nation. Apart from SonicSG we have developed technologies to support urban and interactive media designs that blur the boundary between people, objects, and environments (iSwarm [19] (Fig. 11), nZwarm [20], ReadBridge [21]).

Fig. 11 SonicSG [3] prototype

3 Moving Forward

Our endeavours along different research avenues of Assistive Augmentation were summarized in the chapter. The illustrated research thrusts and application domains correspond to our vision at the Augmented Human Lab: enhancing how we live, work and play and most importantly, humanizing technology. This ranges from practical behavioral issues, understanding real-life contexts in which technologies function to understanding where technologies cannot only be just exciting or novel, but have a meaningful impact on the way people live.

Augmenting senses—or sensors—is key to this agenda. The highlighted application domains specifically focus on enhancing a particular sensory channel, remapping information from one sensory modality to another and creating new sensing modalities. Moving forward, these exemplary projects contribute not only to specific communities but have the potential for wider outreach. In line with the general agenda of Assistive Augmentation as a research field, the emphasis of these projects is rather on "enabling" than on fixing. This approach opens up their potential to a broader range of applications.

References

1. Withana A, Ransiri S, Kaluarachchi T, Singhabahu C, Shi Y, Elvitigala S, Nanayakkara SC. waveSense: ultra low power gesture sensing based on selective volumetric illumination. In: Proceedings of the 29th annual symposium on user interface software and technology
2. Fan K, Huber J, Nanayakkara S, Inami M (2014) SpiderVision: extending the human field of view for augmented awareness. In: Proceedings of the 5th augmented human international conference. ACM, p 49
3. Nanayakkara SC, Wyse L, Ong SH, Taylor E (2013) Enhancing musical experience for the hearing-impaired using visual and haptic inputs. Hum Comput Interact 28(2):115–160
4. Petry B, Illandara T, Nanayakkara SC (2016) MuSS-Bits: sensor-display blocks for deaf people to explore musical sounds. In: Proceedings of the annual meeting of the australian special interest group for computer human interaction, OzCHI '16. ACM, New York
5. Ranasinghe N, Cheok A, Nakatsu R (2014) Taste+: digitally enhancing taste sensations of food and beverages. In: Proceedings of the ACM international conference on multimedia. ACM, pp 737–738
6. Clark RC, Nguyen F, Sweller J (2011) Efficiency in learning: evidence-based guidelines to manage cognitive load. Wiley
7. Eisma R, Dickinson A, Goodman J, Syme A, Tiwari L, Newell AF (2004) Early user involvement in the development of information technology-related products for older people. Univ Access Inf Soc 3(2):131–140
8. Shilkrot R, Huber J, Wong ME, Maes P, Nanayakkara SC (2015) FingerReader: a wearable device to explore printed text on the go. In: Proceedings of the 33rd annual SIGCHI conference on human factors in computing systems, Seoul, Korea, April 18–23, 2015. CHI'15. ACM, New York, NY, pp 2363–2372
9. Cortes JPF, Ching TH, Wu C, Chionh CY, Nanayakkara SC, Foong S (2015) BWard: an optical approach for reliable in-situ early blood leakage detection at catheter extraction points. In: Proceedings of the 7th IEEE international conference on automation and mechatronics (RAM), Angkor Wat, Cambodia, July 15–17, 2015, CIS-RAM 2015. IEEE, Piscataway, NJ
10. Sato M, Poupyrev I, Harrison C (2012) Touche´: enhancing touch interaction on humans, screens, liquids, and everyday objects. In: Proceedings of CHI 2012. ACM Press, pp 483–492. http://doi.org/10.1145/2207676.2207743
11. Yeo KP, Nanayakkara SC, Ransiri S (2013) StickEar: making everyday objects respond to sound. In: Proceedings of the ACM symposium on user interface software and technology, StAndrews, UK, October 8–11, 2013. UIST'13. ACM, New York, NY, pp 221–226
12. Ransiri S, Nanayakkara SC (2012) WatchMe: wrist-worn interface that makes remote monitoring seamless, ASSETS
13. Zhang M, Sawchuk AA (2009) A customizable framework of body area sensor network for rehabilitation. Applied sciences in biomedical and communication technologies, 2009. Isabel
14. Hiroki S, Takashi W, Achmad A (2009) Ankle and knee joint angle measurements during gait with wearable sensor system for rehabilitation. In: World congress on medical physics and biomedical engineering
15. Peiris R, Janaka N, De Silva DR, Nanayakkara SC (2014) SHRUG: stroke haptic rehabilitation using gaming interfaces. In: Proceedings of the 25th annual CHISGI Australian computer-human interaction conference, Sydney, Australia, December 2–5, 2014. OZCHI'14. ACM, New York, NY, pp 380–383
16. Picard RW, Papert S, Bender W, Blumberg B, Breazeal C, Cavallo D, Strohecker C et al (2004) Affective learning—a manifesto. BT Technol J 22(4):253–269
17. Ortega-Avila S, Huber J, Nanayakkarawasam J, Withana A, Fernando P, Nanayakkara SC (2014) SparKubes: exploring the interplay between digital and physical spaces with minimalistic interfaces. In: OzCHI '14: Proceedings of the 26th Australian computer-human interaction conference. ACM

18. Sridhar PK, Nanayakkara SC, Huber J (2017) Towards understanding of play with augmented toys. In: Proceedings of the 8th augmented human international conference, AH '17. ACM, New York, NY, USA. Article 22

19. Nanayakkara SC, Schroepfer T, Wortmann T, Yeo KP, Khew YN, Lian A, Cornelius A (2014) iSwarm: an iterative light installation on the water. In: i Light Marina Bay 2014, Marina Bay, Singapore, 7–30 March, 2014

20. Nanayakkara SC, Schroepfer T, Withana A, Wortmann T, Pablo J (2014) nZwarm: a swarm of luminous sea creatures that interact with passers-by. In: Wellington LUX 2014, Wellington Waterfront, New Zealand, 22–31 August, 2014

21. Nanayakkara SC, Schroepfer T, Withana A, Lian A, Boldu R, Muthukumarana S (2015) The RIBbon @ read bridge: interactive installation that movements and sounds into beautiful light. In: Christmas by the River, Read Bridge, Clarke Quay, Singapore, 4 December 2015–4 January 2016)

22. Nanayakkara SC, Schroepfer T, Withana A, Lian A (2015) SonicSG: interactive mobile-based light installation that tinkles in symphony with the nation's well wishes. In: SG50 celebrations, Boat Quay, Singapore, 4 December 2015–4 January 2016

23. Bishop C (2012) Artificial hells: participatory art and the [6] politics of spectatorship. Verso Books

Scaffolding the Music Listening and Music Making Experience for the Deaf

Benjamin Petry, Jochen Huber and Suranga Nanayakkara

1 Introduction

Music is an important part of our daily life. We listen to the radio, enjoy concerts or make music. However, 360 million people worldwide have limited access to music due to "disabling hearing loss" [63]. Deaf musicians, such as percussionist Evelyn Glennie [16] and opera singer Janine Roebuck [55], are extraordinary examples of individuals who made music their occupation. Initiatives such as "Music and the Deaf" [44], aim to encourage deaf people to make music through seminars, concerts and workshops. Moreover, educational approaches towards making music for deaf people have been investigated [18, 19, 38, 40] and assistive technology for music making such as MOGAT [67] and the Tac-Tile Sound System [50] have developed additional visual and vibrotactile feedback to compensate for the lack of hearing.

In this chapter we describe two assistive augmentations based on sensory substitution techniques for deaf people in the application areas of music listening and music making. Firstly, we discuss and compare existing assistive technologies that support musical activities for deaf people through augmentation of the ear or applying sensory substitution, in particular using visual and vibrotactile feedback. Secondly, we introduce the Haptic Chair system, a chair that provides audio information through a visual display and vibrations transmitted to different body sites, including hands, back and feet. An evaluation of the Haptic Chair shows that

B. Petry (✉) · S. Nanayakkara
Augmented Human Lab, Singapore University of Technology and Design, Singapore, Singapore
e-mail: bpetry@acm.org

S. Nanayakkara
e-mail: suranga@ahlab.org

J. Huber
Synaptics, Zug, Switzerland
e-mail: jochen.huber@synaptics.com

© Springer Nature Singapore Pte Ltd. 2018
J. Huber et al. (eds.), *Assistive Augmentation*, Cognitive Science and Technology, https://doi.org/10.1007/978-981-10-6404-3_3

deaf people enjoy the vibrotactile music listening experience with the Haptic Chair. Additionally, it has been found that the feedback from the Haptic Chair could improve the regular speech therapy program. The third section transits from music listening to music making and introduces the VibroBelt, a reprogrammable music-to-vibrotactile belt worn around the waist. We discuss requirements for music making assistive technology based on an observational study, describe the technical development of the VibroBelt and present user sessions that indicate vibrotactile feedback as a potential modality for scaffolding the music making process for deaf people. We close this chapter with a conclusion and outlook on how we envision the future development of assistive augmentation technology that supports musical activities for the deaf community.

2 Related Work

2.1 Augmentation of the Existing Pathway

Hearing aids are one of the most common strategies for deaf people to experience sound. Hearing aids amplify sound and improve language recognition. However, they are known to distort music which limits its enjoyment [4, 11]. Cochlear implants are another way to augment the ear. A cochlear implant consists of an array of electrodes that is inserted directly into the ear's cochlea to stimulate the nerves with electrical signals. These signals are quite different from natural auditory signals. Hence, the success of a cochlear implant depends on the age of implantation [6]. The earlier a person gets the cochlear implant, the higher is the chance of success. Cochlear implants improve language recognition, however, they are less ideal for music perception [9]. Due to the limited bandwidth of cochlear implants, the perception and identification of melodies [13] and timbre [9] are poor, limiting the musical experience [33]. In addition, cochlear implants require an expensive surgery and can cause ear infections.

2.2 Augmentation of Alternative Sensory Channels

"*Sensory substitution devices (SSDs) convey information that is normally perceived by one sense, using an alternative sense*" [32]. Sensory substitution has been applied to compensate hearing loss for sound type detection [20, 37], direction cueing [24, 60], speech processing using the Tadoma method [54], and the enhancement of musical experiences [26, 28, 49]. The following section provides an overview and comparison of two different sensory substitution strategies for music.

2.2.1 Music-to-Visual Sensory Substitution Strategies

Mori et al. used animated lyrics to convey emotions of music [42, 53, 62]. In fact, signing the lyrics of a song by sign language is a common practice in the deaf community to translate sung music into visual form. Most music-to-visual sensory substitution strategies focus on representing musical information as animations on a screen. An overview is given in Table 1 (for more details see [51]). Common mapping for pitch is the vertical position of an object representing a note (e.g., circle or bar), its color or its angle relative to a reference point, on a screen. However, different instruments are also presented through different colors, which can interfere with the pitch mapping. Time is presented in two different ways: (1) along an axis, which makes past and future of the music visible or (2) as instantaneous events. Given the importance of real-time information, music making approaches mainly use instantaneous events for time representation. To our knowledge, an optimal mapping from musical information to animations has not been established yet. However, Fourney et al. found that visualizations with too much musical information tend to be boring for hearing impaired individuals [12]. Rather, the entertainment value seems more important for deaf music consumers.

2.2.2 Music-to-Vibrotactile Sensory Substitution Strategies

An overview of existing music-to-vibrotactile sensory substitution systems is given in Table 2 (for more details see [51]). We note that the use of pure audio signals is quite common for music-to-vibrotactile sensory substitution approaches. This is reasonable since the audio signal can be directly used as the driving signal for many actuators, such as speakers and voice coils. With this mechanism, pitch and loudness is directly translated to vibrotactile frequency and intensity, respectively. However, the frequency discrimination of the skin is poor compared to audio and the skin's frequency range is small (up to 1000 Hz) [61]. This limits the number of distinguishable frequencies for vibrotactile stimuli. Hence, some approaches combine vibration frequency with spatial location (point of feedback on the skin) to improve pitch discrimination. Furthermore, time is mainly presented through instantaneous events. The representation of instruments has not been the main focus of music-to-vibrotactile sensory substitution strategies. Nonetheless, Karam et al. have found, that an additional mapping of instruments conveys more information [28].

The use of vibrotactile feedback as a substitution strategy for auditory feedback can be challenging, since the properties and limitations of auditory and vibrotactile information are different: the ear can perceive frequencies between 20 Hz and 20 kHz, but the skin's perceptual frequency range is limited to 1000 Hz with a peak frequency at 250 Hz [5, 35, 59, 61]. Further, the spatial tactile resolution varies across different body parts. Some body parts, such as hands or lips, have a very high spatial tactile resolution (discrimination of a two point stimuli of less than 10 mm in average), but other body parts, such as calf or back, have a low resolution (larger

Table 1 Overview of visual sensory substitution strategies for music

Work	Focus (music-listening/music-making)	Audio source	Mapping (selected musical elements)			
			Pitch	Loudness	Instrument	Time
Piano roll view [12, 22, 43][a]	Listening	MIDI	Vertical position		Color	Horizontal position
Part motion view [12, 43][a]	Listening	MIDI	Vertical position		Color	Horizontal position
Tonal compass view [12, 43][a]	Listening	MIDI	Angle	Circle's size		Instantaneous
Motion pixels of music [12][a]	Listening	MIDI	Angle		Color, screen position	In-/outwards movements
MusicViz [52][a]	Listening	MIDI	Vertical position	Pipe's size + brightness	Color/shape[b]	Depth position[b]
MOGAT [67][a]	Making	Audio	Vertical position			Instantaneous
Movies from music [41]	Listening	N/A	Color + brightness			Distance from center
Seen music [30][a]	Making	[c]	Color			Instantaneous
Spectrogram [22]	Listening	Audio	Vertical position			Horizontal position
CAMLS for hearing-impaired [65][a]	Making	Audio	Written text, notation			Position inside the notation
Seeing sound [10]	Making	Audio	Angle	Vertical height		Instantaneous

[a]Designed or evaluated with deaf or hard of hearing individuals
[b]Uses a three-dimensional representation
[c]Exact pitch information is retrieved using an external trigger

Table 2 Overview of vibrotactile sensory substitution strategies for music

Work	Focus (music-listening/music-making)	Audio source	Actuator used	Mapping (selected musical elements)			
				Pitch	Loudness	Instrument	Time
MUVIB [31][a]	Listening	Audio	ERM motor		Intensity		Instantaneous
Tactilicious flute display [1, 36]	Listening	Audio	Voice coil	Frequency	Intensity		Instantaneous
Tac-tile sounds [50][a]	Listening + making	Audio	Speaker	Frequency	Intensity		Instantaneous
Emoti-chair (frequency model) [25–27][a]	Listening	Audio	Voice coil	Frequency + spatial location	Intensity		Instantaneous
Emoti-chair (track model) [25–27][a]	Listening	Audio (separated instrument tracks)	Voice coil	Frequency + spatial location relative to instrument	Intensity	Spatial location[a]	Instantaneous
Emoti-chair (control model) [25–27][a]	Listening	Audio	Voice coil	Frequency	Intensity		Instantaneous
Vibrochord [2]	Making	MIDI	Voice coil	Spatial location + frequency	Intensity		Instantaneous
Mobile music touch [21]	Making	MIDI	ERM motor	Spatial location (finger)			Instantaneous

[a]Designed or evaluated with deaf or hard of hearing individuals

than 40 mm in average) [17]. However, these tests have been conducted with small single point actuators. Using larger area of stimulation could take the advantage of the skin's spatial summation and hence, could result in a better discrimination as proposed by Branje et al. [3] and Wyse et al. [64].

3 Music Listening: The Haptic Chair

In this subchapter, we present the Haptic Chair, a music-to-vibrotactile and music-to-visual sensory substitution system for listening. Technical details of the development of a vibrotactile and a visual display are discussed along with an evaluation with profoundly deaf children. The study provides evidences for the Haptic Chair's effectiveness for listening to music and its potential as an extension to speech therapy.

3.1 Motivation

The motivation behind the Haptic Chair was to enhance the experience of listening to music for deaf people. Deafness does not prevent a person from enjoying music. However, the interest displayed by a person with hearing impairment in music, depends on his or her affiliation with the hearing or deaf culture [8]. The Haptic Chair is inspired by Evelyn Glennie, a profoundly deaf musician, who wrote in her hearing essay: *"Hearing is basically a specialized form of touch"* [15]. She argues that sound is basically vibrating air molecules which are picked-up by the ear. However, sometimes the skin also picks-up these air-conducted vibrations, as they are for low frequency sounds. The idea behind the Haptic Chair is to amplify these vibrations and deliver them to different parts of the body. Evelyn Glennie also mentioned that she discriminates sound according to the location along her body where she feels the sound [16]. Palmer developed a theory that describes that low-middle- and high-frequencies can be felt in different parts of the body [48], which is consistent with the review on the tactile modality, carried out by the Army Research Laboratory, USA [45]. This suggests, that the body is able to pick-up and discriminate sound based on the different body parts through vibrations.

3.2 The Haptic Chair

The Haptic Chair consists of a densely laminated wooden chair that is widely available at relatively low cost (Poäng made by IKEA). Contact speakers (SolidDrive SD1 and Nimzy Vibro Max) were attached to four positions of the chair: one under each armrest, one under a rigid, laminated wooden footrest (also

Poäng by IKEA) that was securely fixed to the main chair, and one on the backrest at the level of the lumbar spine. To improve the contact of the hand with the chair, one hand doom at each side was mounted at the appropriate positions to the chair. An amplifier (SD-250 mini amplifier) takes the sound input from a sound source (e.g. mobile phone or computer), amplifies it and distributes it to the four contact speakers. The mechanism of providing a tactile sensation through this wooden structure is quite similar to the common technique used by deaf people, called "speaker listening" where deaf people place their hands or foot directly on an audio speaker. However, the Haptic Chair provides a tactile stimulation to various parts of the body simultaneously in contrast to speaker listening, where only one part of the body is stimulated at any particular instant and not necessarily within an optimal frequency range. A complete sketch of the Haptic Chair is shown in Fig. 1.

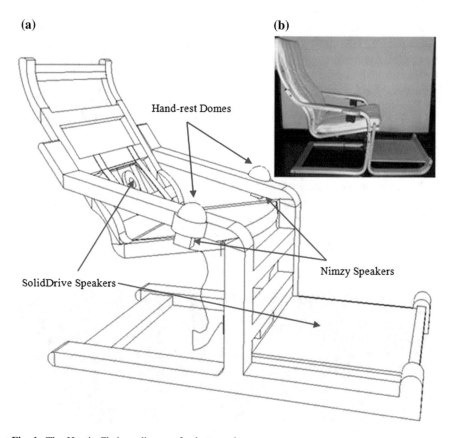

(a)

(b)

Hand-rest Domes

Nimzy Speakers

SolidDrive Speakers

Fig. 1 The Haptic Chair: **a** diagram. **b** photograph

3.3 Visual Display

A visual display was developed to investigate the effect of additional visual feed-back. A fundamental decision in designing a music-to-vision display is the type of visualization related to time: piano roll or movie roll (see Fig. 2). The piano roll presentation refers to a display that scrolls from left to right. The axis of scrolling represents time. Musical events occurring at a specific time are displayed in a single column. This enables the user to see the past, current and future of events. How-ever, this does not relate to hearing, where the listener cannot hear future sounds neither past sounds, except from auditory memory. In contrast, the movie roll presentation shows only instantaneous events allowing more freedom of expression. The visual effect for a particular audio feature is visible on screen for as long as that audio feature is audible and fades away into the screen as the audio feature fades. This is closer to how humans hear sound events and was therefore chosen as the visualization type.

A further challenge was the design of the mapping from the auditory domain into visual effects. As a basic shape a sphere was chosen for each note produced by a non-percussive instrument. With the feedback of two deaf musicians (a pianist and a percussionist) the following auditory parameters were mapped to visual param-eters changing the sphere's appearance:

- **pitch of a note**: vertical position (high pitch—top, low pitch—bottom) and size (high pitch—small, low pitch—big)
- **loudness**: brightness
- **instrument timbre**: color

Figure 3 shows a screen capture of the visual display but obviously cannot convey the effect of a visual display corresponding to a piece of music: this must be left to the imagination of the reader.

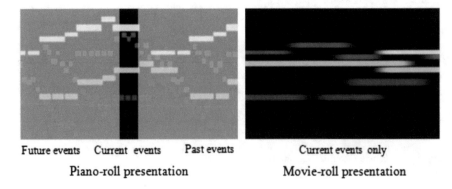

Future events	Current events	Past events	Current events only
Piano-roll presentation			Movie-roll presentation

Fig. 2 Examples of piano-roll and movie-roll presentation

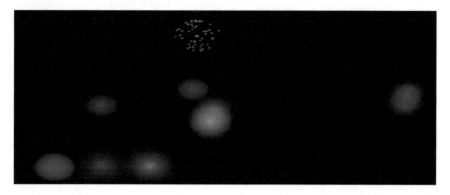

Fig. 3 Screen capture of the visual display with real-time abstract animations corresponding to music

3.4 Evaluation

We conducted a study with 43 hearing-impaired children (28 partially deaf, 15 profoundly deaf—see Fig. 4) ranging from 12 to 20 years to investigate the following questions:

1. Does the visual display enhance their experience?
2. Does the haptic display enhance their experience?
3. Does a combined output (visual and haptic display) enhances their experience?
4. What is an optimal configuration?—Visual display alone, haptic display alone, visual and haptic display together.

To evaluate the experience of a participant, we used Csikszentmihalyi's theory of flow [7]. Csikszentmihalyi describes "being in the flow" as the timelessness, effortlessness, and lack of self-consciousness one experiences. He described "flow" as a state in which people are so involved in an activity that nothing else matters: The experience itself is so enjoyable that people will do it even at a high

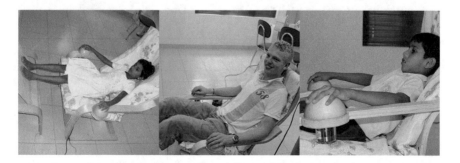

Fig. 4 Deaf children using the Haptic Chair

cost, for the sheer joy of doing it. To measure the flow, a questionnaire called the Flow State Scale (FSS) with 36 items was used [23].

In the study, the Haptic Chair was used as the haptic display and a 17″ LCD display for the visual effects. In addition, a normal diaphragm speaker system (Creative 5.1 Sound Blast System) was used to play the music. The experiment was a within-subject 4 × 3 factorial design. The two independent variables were music sample (classical, rock, beat only—each was 1 min in length) and the test condition (music only, music + visual display, music + haptic display, music + visual + haptic display). The samples were presented in random order. The task throughout the study was to follow the music on the activated displays.

3.5 Discussion

The FSS score was minimal when only the music was played and the visual and haptic display were turned off. Furthermore, there was no main effect for music genres and the level of deafness as well as no interaction between music genre and test condition. However, there was a main effect between the four test-conditions as shown in Fig. 5. The "music only" condition was significantly smaller than the other 3 conditions (p < 0.01). Furthermore, the "music + haptic" as well the "music + visual + haptic" conditions' scores were significantly higher than the "music + visual" score (p < 0.01). When the participants were asked, which combination they preferred most, 54% preferred the "music + haptic" setup, 46% the "music + visual + haptic" setup and no participant one of the other two options. These findings suggest that the haptic display has a high contribution to the enhancement of the music listening experience.

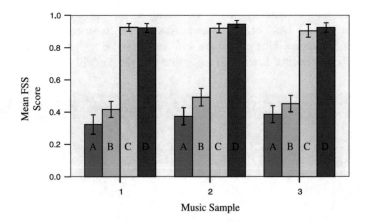

Fig. 5 Overall flow state scale (FSS) score for three music samples under all experimental conditions with error bars showing 95% confident interval (*A*—music alone, *B*—music and visual display, *C*—music and haptic display, *D*—music, visual and haptic display)

The current system, makes no attempt to electronically process speech in any way, but instead delivers the entire input audio stream to each of the separate vibration systems targeting the feet, back, arms and hands. This is not necessarily the optimal strategy for vibrotactile presentation [26–28]. In future work, we will explore the possibility of providing customized (e.g. separated by frequency bands) vibrotactile feedback through different vibration elements to different locations on the body. Moreover, we are focusing on extending the Haptic Chair concept into a wearable or portable device. We hope that these future works will lead to more effective uses of the vibrotactile channel for music as well as communication via speech for the profoundly deaf.

3.6 Extending to Speech Therapy

Prolonged use of single-user Haptic Chair unveiled the potential to be a useful tool in speech therapy—going beyond the original aim of enhancing the pleasure of 'listening to music'. In a typical speech therapy session at the school, a deaf student and a speech therapist sit in front of a mirror. The student watches the speech therapist's lip movement in the mirror and tries to mimic those movements. We observed that the students are often able to mimic lip movements, but either they generate no sound or they generate a sound that is very different from the example provided by the therapist. This is not surprising given the lack of auditory feedback. Furthermore, it was also clear that many profoundly deaf students did not enjoy the speech therapy sessions, which is a common problem worldwide.

Almost a century ago, Gault [14] proposed a method of presenting speech signals via a vibrator placed on the skin. This provided further motivation for exploring this kind of vibrotactile feedback for speech therapy and education. The design of the Haptic Chair was extended so that users would be able to sense amplified vibrations produced by their own voice as well as others such as teachers or therapists. With this modification, we observed immediate effects on the awareness the profoundly deaf users had and whether they were matching the sound production pattern accompanying the lip movements they could see.

We conducted a 12-week long pilot study and a 24-week long follow-up study to evaluate the effectiveness of the Haptic Chair in speech therapy sessions for profoundly deaf students (see Figs. 6 and 7). Our results suggest that this kind of display can, to some extent, function as an effective substitute for the traditional 'Tadoma' [54] method of speech instruction wherein students touch the throat or lips of their teachers. It would also open up a range of approaches for speech therapy aids that are independent of or complementary to the physical presence of a human therapist.

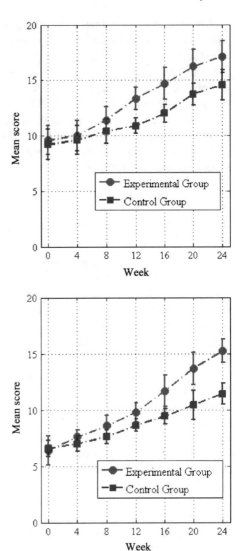

Fig. 6 Follow-up study: mean score (average number of words recognized by the **speech therapist**) after every 4-weeks with 95% confidence interval

Fig. 7 Follow-up study: mean score (average number of words recognized by the **independent listener**) after every 4-week block with 95% confidence interval

4 Music Making: The VibroBelt

The previous sub-chapter provided a study of enhancing the music listening experience using a visual and a haptic display. In this subchapter, we focus on music making. We introduce the VibroBelt, a music-to-vibrotactile sensory sub-stitution system to support deaf people in music making. We first conducted an observational study to understand the needs and requirements for music making systems for profoundly deaf children. We present the implementation of a wearable haptic display for exploring ways of scaffolding the music making process for deaf people and provide initial user reactions.

4.1 Motivation

Music making differs from music listening as it is *"more powerful [...], transformative [...], and a way to express yourself"* [34]. Furthermore, it depends on a *"strong coupling of perception and action mediated by sensory, motor and multimodal brain regions"* [56, 66]. Hence, music making requires a closed feedback loop (play-listen-evaluate-correct-play) that allows the control of musical elements, such as pitch, melody and rhythm. This can be challenging for people with hearing impairments for musical elements that mainly depend on tonal content, such as pitch and melody. These elements are almost impossible for them to perceive and tonal instruments, such as the piano and guitar, are therefore harder to play.

4.2 Need and Requirement Analysis

We conducted an observational study with 7 congenitally profoundly deaf children aging from 12 to 15 years. The aim of this study was to understand needs and requirements for music making. In particular, we were interested in the following questions:

1. How do profoundly deaf children interact with musical instruments?
2. Which musical elements do profoundly deaf children vary?
3. Which strategies do profoundly deaf children employ to compensate for their hearing loss?

The study had two parts: (A) an interview and (B) a musical activity. The latter consisted of two phases: (1) instrument exploration and (2) musical expression. The semi-structured interview focused on the children's prior experience with music and musical instruments. Following that, each child selected one instrument out of 5 for the musical activity. We brought 3 non-tonal instruments (Shakers, Thammettama Drum, and Bass Drum) as well as 2 tonal instruments (Guitar and Violin) as shown in Fig. 8. In the exploration phase, each child could familiarize himself or herself with the instrument. Initially, no instructions were given; but instructions were given if a child did not know how to interact with the instrument. In the musical expression phase, we asked the children to use instruments to express familiar

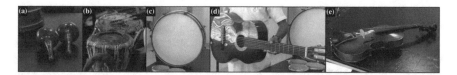

Fig. 8 The instruments we used: **a** shakers, **b** thammettama drum, **c** bass drum, **d** guitar and **e** violin

phenomena: (1) happiness and sadness, (2) a running rabbit and a crawling turtle and (3) a bird flying up to the sky and rain falling down to the ground.

The duration of each session was about 45 min. Each child tried 2–3 different instruments. The musical activity was video recorded for further analysis. Furthermore, 3 researchers observed and took written notes throughout the whole study. The data was coded iteratively after each session, resulting in 6 main categories.

4.2.1 Prior Experience

All children reported to have experience with non-tonal instruments, such as Cymbal, Thammettama Drum and Bass Drum. Furthermore, all children stated to enjoy playing those instruments. However, the children could not specifically say, what they like about playing the instruments: *"[I] feel happy when playing" (P1)*, *"[It] makes [me] not sad, when [I am] playing" (P2)*.

4.2.2 Challenges of Playing Instruments

Challenges were extracted through observations during the musical activity part. Some children reported to have doubts whether they played the instrument correctly: *"[I don't] know, if [I] did it correctly" (P5), others had no doubts: "[I] knew, [I] did it" (P6)*. It seems, that some of the children care about what others, especially hearing people, may hear and are aware that they perceive more than them. On the other hand, some of the children focus on their own experience. As long as they enjoy this experience, there is no challenge.

4.2.3 Feedback Strategies

The deaf children showed several strategies to compensate for their hearing loss. We found two main visual feedback strategies: (1) focusing (see Fig. 9a) and (2) looking at the audience. A very strong feedback strategy employed by most of the children is to focus on the instrument and the area where the action happens, e.g. bow touches the strings for violin, fingers touching the strings on a guitar. This strategy was also reflected by the participants' feedback: *"[I am] seeing what [I am] doing" (P4)*. In addition, P4 also uses the audience's reaction to verify if she is performing correctly: *"Other people's reaction to what [I] do confirms, what [I am] doing is right"*.

A further strategy to get feedback from the instrument is to perceive its natural tactile feedback. We expected this to be a main strategy for them. However, only P2 mentioned to feel the Violin in his left arm and P6 said *"[I don't] know [where I felt*

Fig. 9 Pictures from the observational study: **a** participant is visually focused on the instrument, **b** participant replies to questions and **c** participant imitates finger movements on the guitar's neck while visually focusing on his action

the sound]." Furthermore, they used the feedback of their own motor movement to verify if they perform properly: *"[I] felt the [faster] movement of my hand" (P6)*.

4.2.4 Feedback Interpretation

During the sessions the children often referred to hearing (as shown in Fig. 9b): *"[I] hear, but [I don't] know what [I] hear" (P1)*. Even when we specifically asked, whether they *heard* or *felt* something on the skin, they referred to hearing: *"[I] didn't feel something in [my] hand, but [I] heard something" (P7)*. Furthermore, P7 mentioned to hear vibrations: *"The vibrations of the [guitar's] string, that's what [I] heard"*. Based on these comments, we assume, that they refer to hearing as the result of processed visual and tactile feedback.

4.2.5 Playing Strategies

During their play, we observed two main playing strategies: (1) counting and (2) mimicking (see Fig. 9c). We observed, that the children quickly converged to a regular rhythm and did not vary any musical elements from that point onwards. To keep a steady beat they counted: *"[I] was counting: 3 and 1" (P1)*. Moreover, we asked them to play without counting and saw that some children struggled. However, we also observed that they finally converged again to a different counting pattern. Another strategy we saw was mimicking. When we observed that they used their fingers on the Violin's or Guitar's neck, they said: *"[I saw] other people doing [it]" (P3)*, *"[I] used [my] fingers, because [I] saw it on TV" (P2)*. However, it would be interesting to see, if the strategies of play would change or further ones would arise when additional feedback is provided.

4.2.6 Musical Interpretation

Music making offers the possibility for self-expression. We were interested which musical elements profoundly deaf children can vary to express familiar phenomena.

Thus, the musical expression task gave us insights in the children's ability of varying musical elements. As an initial step, we focused on 3 different musical elements: Tempo, Loudness and Pitch. Happy and sad was mainly expressed through tempo and loudness. This was also the intent of the participants as they stated: *"Happy louder, sad softer" (P4)*. However, for P2 it was the opposite way when he was playing the guitar: *"When happy, [I] slowly play the guitar. When sad, [I] played fastly."* While most participants could exactly say what they changed in their play to express happiness or sadness, P2 and P7 applied another strategy: *"When [I] played, [I] imagined to feel sad" (P2)*, *"For happiness [I] thought on [my] mother and played something" (P7)*.

Since we saw only tempo and loudness variations for the interpretation of happy and sad for P1–P4, we introduced two new phenomena with clear contrast for P5–P7. The interpretation of a running rabbit and a crawling turtle were aiming for tempo variations only. The participants expressed the rabbit as fast throughout the experiment while the turtle was expressed as slow. This was also reflected by the participants' comments: *"[I will] play fast and slow" (P6)*. The interpretation of the picture of a bird flying up to the sky and rain falling down from the clouds were aiming for tonal variations. However, all participants expressed those pictures with tempo variations. For the bird the participants mostly started slow and got faster in tempo: *"Bird starts slow and accelerates [as it] goes up" (P5)*. For the rain, it was mostly the opposite way. However, P5 interpreted rain falling down as slow, then getting faster and becoming slow again. She commented that, *"Rain comes down and slowly floats away"*.

4.3   Opportunities for Scaffolding Music Making

4.3.1   Tonal Variations

Profoundly deaf children have the ability to express familiar phenomena with instruments as we reported. However, their dimension of expression was limited to tempo and loudness. We did not see any tonal variations even on tonal instruments throughout the whole study. This could be due to two main reasons: (1) they do not know that tonal variation exists since they are deaf or (2) they know tonal variation exists, but they do not know how to apply it on the instrument. We believe it is mainly due to the first reason; for example, we saw no effort in changing squeaking sounds that P1, P2 and P3 accidentally produced with the violin. Nonetheless, it seems that improving the feedback loop for tonal feedback, such as pitch or melody, has a lot of potential to enhance their music making experience.

4.3.2 Visual Feedback

Visually focusing on the area of action is a strong strategy to receive feedback from the instrument. Visual feedback as a sensory substitution strategy can interfere with or complement their current strategy. We speculate that visuals on a screen, as it has been done in prior work for music listening (e.g. [12, 22, 52]), interferes with their focusing strategy. However, enhancing the area of action with additional feedback (e.g. lighting up the keys of a keyboard as in [39]) could complement their focusing strategy. In summary, it is important to minimize attention switching.

4.3.3 Vibrotactile Feedback

Prior work has shown, that vibrotactile feedback is an effective channel for music making [15, 40, 50]. In our study, we did not see our participants relying much on that channel. However, we think that vibrotactile feedback is still an important channel for music making, since previous studies [29, 57] found that some deaf people process tactile sensations in parts of the brain that are associated with hearing in hearing people. Furthermore, as discussed in the previous subchapter, the haptic display of the Haptic Chair has a high contribution to the music listening experience at least.

4.3.4 Other Considerations

Music making requires free limb movement. Since most instruments are played with hands or feet they should not be restricted in any way. Using limbs for the perception of feedback, such as vibrotactile feedback on the fingertips, could be distracting and disrupting the music making experience. Furthermore, music making assistive technology should be comfortable, since music making is an activity that can range from minutes to hours.

4.4 The VibroBelt

Based on prior work regarding music making for deaf people [30, 65, 67] and the results of our observational study, we decided to focus on conveying tonal information. To avoid interference with the children's visual focusing strategy, we decided to use a music-to-vibrotactile sensory substitution strategy. We developed a wearable vibrotactile belt (see Fig. 10) that provides vibrations to the human's back. This allows free limb movement and a comfortable use over a longer period of time. Though, the back's sensitivity is not that high, it provides a large surface allowing us to make use of spatially-encoded information. The VibroBelt consists of 8 light weighted coin vibrator motors (ERM motors—Model 310-101 from

Fig. 10 Participant wearing the VibroBelt. The prototype consists of 8 stripes with attached vibrator motors

Precision Microdrive). Each motor is attached to a 200 × 16 × 2 mm acrylic stripe to enhance vibrations and to allow spatial summation by the skin. The motors are connected via a 1 m long cable to the VibroBelt-Controller, consisting of a multi driver board (DRV2605L Multi-Driver Board from Texas Instruments) and an Arduino Mini Pro (3.3 V). The motor-stripes were attached to a commercially available belly-belt in a horizontal arrangement with a distance of 20 mm between two stripes. We connected the VibroBelt with a keyboard (Yamaha SY22 MIDI-keyboard with semi-weighted keys) to understand if vibrotactile feedback allows profoundly deaf children to make tonal variations. An overview of the system is shown in Fig. 11.

4.5 Evaluation

We conducted 3 user sessions with 4 profoundly deaf children to investigate whether vibrotactile feedback has the potential to scaffold the music making process. In particular, we were interested whether the VibroBelt supports

- discrimination of tones
- identification of tones and
- reproduction of tone sequences.

4.5.1 Setup

We connected the VibroBelt to a keyboard and used it as an initial prototype. Since the children had no experience with a keyboard so far, we only used 8 white keys (C4–C5) to simplify the tasks. Furthermore, we implemented three music-to-vibrotactile mappings:

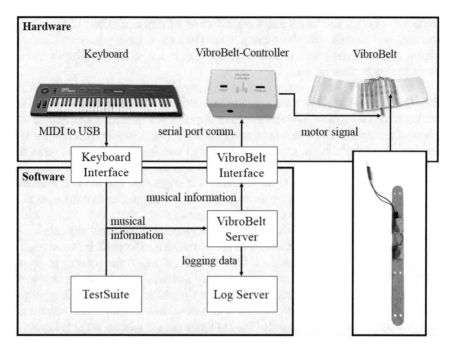

Fig. 11 The VibroBelt system

- Mapping A: Inspired by the Emoti-Chair [25, 28], each of the 8 keys is spatially mapped to one of the 8 motors on the VibroBelt.
- Mapping B: This mapping activates all motors at the same time with a "tapping"-pattern related to the key's representative frequency.
- Mapping C: Inspired by David Eagleman's work [47], this mapping uses left-to-right moving spatiotemporal patterns related to the key's representative frequency.

For mapping B and C the key's original frequencies (C4–C5: 261.63–523.25 Hz) were shifted down to a tactile perceivable frequency range (C0–C1: 16.35–32.70 Hz).

4.5.2 Stimuli and Procedure

User Session 1—Tone Discrimination: Two tones (each 1000 ms in duration) with a break of 1000 ms were presented through the VibroBelt to the participant's back. The stimuli set consisted of 8 same-tone pairs (e.g. C4 and C4) and 7 adjacent-tone pairs (e.g. C4 and D4). Each pair was tested 3 times and the order of tone-pairs was randomized. Each stimulus was played from a computer and no visual cues were available to the participant. After the participant perceived a tone-pair, he or she had to decide whether the two tones were the same or different. Their response was recorded and the next tone-pair was played. For each mapping the whole set was played before moving to the next mapping. The order of the mappings within a participant was randomized.

User Session 2—Tone Identification: The stimuli set consisted of 8 tones (C4–C5). Each tone was played for a period of 1000 ms. The order of the tones was randomized. Throughout the whole session mapping A was used. The stimulus was played from the computer and no visual cues were given. Before a session started, the participant had time to explore the keyboard wearing the VibroBelt. After the participant perceived a tone, he or she tried to find the tone on the keyboard. The participant reported when he or she had found the right tone and the next tone was played. The participant could try as long as he or she wanted, to find the right tone and could ask the experimenter to play the tone again.

User Session 3—Reproduction of Tone Sequences: Short tone sequences (between 4 and 8 tones) were presented to the participants. The participant was asked to find and repeat the sequence on the keyboard. To ease the process of reproduction, the first tone of all sequences was C4 and was marked on the keyboard. The stimulus was presented from the computer without any visual cues. Mapping A was used throughout the whole session. The session started with 4 tone sequences. The participant was given unlimited time to figure out the sequence on the keyboard and could ask the experimenter to play the sequence again. After the participant indicated that he or she had found the sequence, he or she was asked to play the sequence again. The response (including the exploration) was recorded on a computer. The same procedure was repeated for 5-, 6-, 7-, and 8-tone sequences. If the task became too difficult, the participant had the option to drop out.

4.5.3 Results

User Session 1—Tone Discrimination: We found that participants performed above chance for mapping A, but around chance for mapping B and C (see Fig. 12). To get deeper insights in the performance of mapping B and C, a confusion table was created. It revealed that 79% of all responses were *"same"* for mapping C indicating that participants perceived most tone pairs as same, independently of a same- or different-tone-pair stimulus. Based on these result we used mapping A for the following sessions, since these require the ability to discriminate.

User Session 2—Tone Identification: We observed, that the participants were able to find the right key in at least 50% of the trials on average (see Fig. 13). Further analysis revealed that if a participant selected a wrong key, he or she would mainly be one key above or below the correct key (93 out of 96 trials). There were 2 instances of selecting a wrong key that was 2 keys apart and 1 instance of selecting a wrong key that was 3 keys apart from the correct key. Even the first guess was observed to be quite close to the correct key.

User Session 3—Reproduction of Tone Sequences: This task was difficult for some participants. Only one participant completed all tone sequences. We analyzed

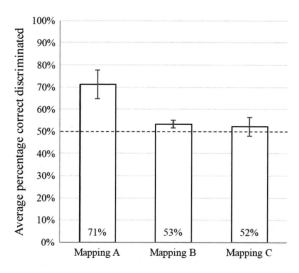

Fig. 12 The average percentage of correctly discriminated tone-pairs across 4 participants (95% confidence interval)

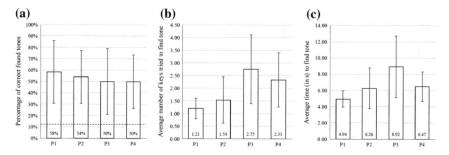

Fig. 13 Tone identification test: **a** percentage of correct identified tones per participant, **b** average number of keys pressed and **c** time needed to find a key (The *dotted line* represents the chance level to hit the right tone)

the recorded melodies of the participants and compared them to the actual presented melodies. Even for hearing people, finding the exact tones of a melody might be difficult. However, getting the contour of a melody (*up* and *down*) right should be possible. Exact tone-hits ranged from 35 to 70%. Regarding identifying the contour of a melody, participants were able to achieve 60–85% of accuracy. This is also in accordance with our second session (tone identification), where exact hits were around 50% and wrong keys mainly one key apart.

4.5.4 Discussion

Mapping A performed better compared to mapping B and C and allowed the children to determine a given key quite accurately. This could be due to the fact that the haptic stimulation had the same spatial arrangement as the keys on the keyboard. In addition, the vibrotactile feedback was perceived as positive by the participants: "[I like] the vibrations" (P2), "When [I am] playing, it vibrates. That is good" (P4). Once the participants got familiar with the instrument, we asked them to interpret the familiar phenomena from the contextual enquiry: (1) happy and sad, (2) a running rabbit and a crawling turtle and (3) a bird flying up to the sky and rain falling down from the clouds. P1 and P4 only varied tempo similar to what they did in first study. In contrast, P2 and P3 used multiple keys simultaneously to express happiness and the running rabbit and only one key at a time for sadness and the crawling turtle. This change in their interpretation could be a first indicator for the effect of scaffolding via vibrotactile feedback. Furthermore, we also observed the visual focusing strategy throughout all participants and all tasks.

4.6 Limitations of the Current System

The motors we used have a lag time (time until the vibrations can be felt) of 47 ms and a rise time (time until the motor reaches half of its maximum intensity) of 91 ms. This limits the range of "tapping"-frequencies that can be presented. The frequency range we choose to operate on (16.35–32.7 Hz) was representable due to summation of the motor's speed. However, this results in a lower intensity contrast. This could be one reason why mapping B and C performed at chance. We are building a second prototype with new motors (Model 307-103 from Precision Microdrive), which have a lag time of 8 ms and rise time of 28 ms. Furthermore, these motors have a higher maximum intensity of 7G compared to the old motors (1.34G). This should improve the vibrotactile contrast.

We aim to investigate new music-to-vibrotactile mappings that are more closely related to music, such as including harmonics or using audio compressed information to mitigate potential cognitive overload. Moreover, other body sites, such as the ear or shoulder, could be also a good place for vibrotactile pitch feedback. Evelyn Glennie mentioned in her Hearing Essay [15] that she discriminates tones via the body sites or organs where she can feel the vibrations most.

5 Conclusion and Outlook

In this chapter, we presented two assistive augmentation approaches for deaf people to experience music. These two systems are one step towards assistive music making augmentation systems. The Haptic Chair uses a vibrotactile and a visual display to enhance the music listening experience and found that the vibrotactile display has a high contribution towards this experience. The VibroBelt uses a wearable vibrotactile display to explore ways of scaffolding the music making process. Both approaches showed the importance of vibrotactile feedback as it provided an intuitive understanding of music for profoundly deaf children.

As discussed in the related work, there are several sensory substitution strategies for the deaf for music listening as well as music making. These approaches focus either on visual or vibrotactile feedback. We also pointed out, that pitch representation in visual and vibrotactile feedback can lead to challenges. Ambiguous visual representations (such as color for pitch and instruments) or the limited number of distinguishable vibrotactile frequencies limits the amount of information conveyed via these channels. A multi-modal assistive augmentation system, embodying both, visual and vibrotactile feedback, could improve the perception of musical elements and has the potential to enhance music listening and improve music making activities.

However, it is important to note that deafness occurs in very different varieties (e.g. unilateral/bilateral, level of deafness, congenital/deafened). Hence, one generalizable solution for music listening or music making cannot be expected. The same has been found by Shinohara and Tenenberg [58] for the design of assistive augmentation systems for blind people. The current approaches for music listening and music making, even developed with hearing impaired users, provide mainly one fixed mapping for every user. We think it is important to design assistive augmentation system that can be calibrated and customized by the user to allow him or her to explore sound in general and music in particular. A deafened user with severe hearing loss, might use vibration feedback differently than a congenital profoundly deaf user. Catering for individual requirements could be the next step towards building assistive music augmentation system for the deaf community.

References

1. Birnbaum DM, Wanderley MM (2007) A systematic approach to musical vibrotactile feedback. In: Proceedings of the international computer music conference (ICMC), pp 397–404
2. Branje C, Fels DI (2014) Playing vibrotactile music: a comparison between the vibrochord and a piano keyboard. Int J Hum Comput Stud 72(4):431–439. http://doi.org/10.1016/j.ijhcs.2014.01.003
3. Branje C, Maksimouski M, Karam M, Fels DI, Russo F (2010) Vibrotactile display of music on the human back. In: Proceedings of the 2010 third international conference on advances in computer-human interactions. IEEE Computer Society, pp 154–159. http://doi.org/10.1109/ACHI.2010.40
4. Chasin M (2003) Music and hearing aids. Hear J 56(7):36. http://doi.org/10.1097/01.HJ.0000292553.60032.c2
5. Cholewiak RW, McGrath C (2006) Vibrotactile targeting in multimodal systems: accuracy and interaction. In: 2006 14th symposium on haptic interfaces for virtual environment and teleoperator systems. IEEE, pp 413–420

6. Cook AM, Polgar JM (2014) Assistive technologies: principles and practice. Elsevier Health Sciences
7. Csikszentmihalyi M (2000) Beyond boredom and anxiety. Jossey-Bass, San Francisco, CA, US
8. Darrow A-A (1993) The role of music in deaf culture: implications for music educators. J Res Music Educ 41(2):93–110. http://doi.org/10.2307/3345402
9. Drennan WR, Rubinstein JT (2008) Music perception in cochlear implant users and its relationship with psychophysical capabilities. J Rehabil Res Dev 45(5):779–789
10. Ferguson S, Moere AV, Cabrera D (2005) Seeing sound: real-time sound visualisation in visual feedback loops used for training musicians. In: Ninth international conference on information visualisation, 2005. Proceedings, pp 97–102. http://doi.org/10.1109/IV.2005.114
11. Fourney D (2012) Can computer representations of music enhance enjoyment for individuals who are hard of hearing? In: Miesenberger K, Karshmer A, Penaz P, Zagler W (eds) Computers helping people with special needs. Springer, Berlin, pp 535–542. http://doi.org/10.1007/978-3-642-31522-0_80. Accessed 9 Feb 2015
12. Fourney DW, Fels DI (2009) Creating access to music through visualization. In: Science and technology for humanity (TIC-STH), 2009 IEEE Toronto international conference, pp 939–944. http://doi.org/10.1109/TIC-STH.2009.5444364
13. Galvin JJ, Q-J Fu, Shannon Robert V (2009) Melodic contour identification and music perception by cochlear implant users. Ann NY Acad Sci 1169(1):518–533. http://doi.org/10.1111/j.1749-6632.2009.04551.x
14. Gault RH (1926) Touch as a substitute for hearing in the interpretation and control of speech. Arch Otolaryngol Head Neck Surg 3(2):121
15. Glennie E (1993) Hearing essay. http://www.evelyn.co.uk/literature.html. Accessed 12 Oct 2014
16. Glennie E (2003) How to truly listen. http://www.ted.com/talks/evelyn_glennie_shows_how_to_listen?language=en. Accessed 4 Dec 2014
17. Goldstein E (2009) Sensation and perception. Cengage Learning
18. Hagedorn VS (1992) Musical learning for hearing impaired children. Research perspectives in music education. http://eric.ed.gov/?id=ED375031. Accessed 16 Feb 2015
19. Hash PM (2003) Teaching instrumental music to deaf and hard of hearing students. Res Issues Music Educ 1(1). http://eric.ed.gov/?id=EJ852403. Accessed 16 Feb 2015
20. Ho-Ching FWL, Mankoff J, Landay JA (2003) Can you see what i hear?: The design and evaluation of a peripheral sound display for the deaf. In: Proceedings of the SIGCHI conference on human factors in computing systems. ACM, pp 161–168. http://doi.org/10.1145/642611.642641
21. Huang K, Starner T, Do E, et al (2010) Mobile music touch: mobile tactile stimulation for passive learning. In: Proceedings of the SIGCHI conference on human factors in computing systems. ACM, pp 791–800. http://doi.org/10.1145/1753326.1753443
22. Isaacson EJ (2005) What you see is what you get: on visualizing music. In: ISMIR. Citeseer, pp 389–395
23. Jackson SA, Marsh HW, et al (1996) Development and validation of a scale to measure optimal experience: the flow state scale. J. Sport Exerc Psychol 18:17–35
24. Jain D, Findlater L, Gilkeson J, et al (2015) Head-mounted display visualizations to support sound awareness for the deaf and hard of hearing. In: Proceedings of the 33rd annual ACM conference on human factors in computing systems. ACM, pp 241–250. http://doi.org/10.1145/2702123.2702393
25. Karam M, Branje C, Nespoli G, Thompson N, Russo FA, Fels DI (2010) The emoti-chair: an interactive tactile music exhibit. In: CHI '10 extended abstracts on human factors in computing systems. ACM, pp 3069–3074. http://doi.org/10.1145/1753846.1753919
26. Karam M, Nespoli G, Russo F, Fels DI (2009) Modelling perceptual elements of music in a vibrotactile display for deaf users: a field study. In: Second international conferences on advances in computer-human interactions, 2009. ACHI '09, pp 249–254. http://doi.org/10.1109/ACHI.2009.64

27. Karam M, Russo F, Branje C, Price E, Fels DI (2008) Towards a model human cochlea: sensory substitution for crossmodal audio-tactile displays. In: Proceedings of graphics interface 2008. Canadian Information Processing Society, pp 267–274. http://dl.acm.org/citation.cfm?id=1375714.1375759. Accessed 5 Feb 2015

28. Karam M, Russo FA, Fels DI (2009) Designing the model human cochlea: an ambient crossmodal audio-tactile display. IEEE Trans Haptics 2(3):160–169. http://doi.org/10.1109/TOH.2009.32

29. Kayser C, Petkov CI, Augath M, Logothetis NK (2005) Integration of touch and sound in auditory cortex. Neuron 48(2):373–384. http://doi.org/10.1016/j.neuron.2005.09.018

30. Kim J, Ananthanarayan S, Yeh T (2015) Seen music: ambient music data visualization for children with hearing impairments. In: Proceedings of the 14th international conference on interaction design and children. ACM, pp 426–429. http://doi.org/10.1145/2771839.2771870

31. La Versa B, Peruzzi I, Diamanti L, Zemolin M (2014) MUVIB: music and vibration. In: Proceedings of the 2014 ACM international symposium on wearable computers: adjunct program. ACM, pp 65–70. http://doi.org/10.1145/2641248.2641267

32. Levy-Tzedek S, Hanassy S, Abboud S, Maidenbaum S, Amedi A (2012) Fast, accurate reaching movements with a visual-to-auditory sensory substitution device. Restorative Neurol Neurosci 30(4):313–323

33. Limb C (2011) Building the musical muscle. http://www.ted.com/talks/charles_limb_building_the_musical_muscle. Accessed 22 Nov 2014

34. Machover T (2008). Tod Machover + Dan Ellsey: Inventing instruments that unlock new music. http://www.ted.com/talks/tod_machover_and_dan_ellsey_play_new_music?language=en. Accessed 22 Nov 2014

35. Mahns DA, Perkins NM, Sahai V, Robinson L, Rowe MJ (2006) Vibrotactile frequency discrimination in human hairy skin. J Neurophysiol 95(3):1442–1450

36. Marshall MT, Wanderley MM (2006) Vibrotactile feedback in digital musical instruments. In: Proceedings of the 2006 conference on new interfaces for musical expression. IRCAM—Centre Pompidou, pp 226–229. http://dl.acm.org/citation.cfm?id=1142215.1142272. Accessed 12 Aug 2015

37. Matthews T, Fong J, Mankoff J (2005) Visualizing non-speech sounds for the deaf. In: Proceedings of the 7th international ACM SIGACCESS conference on computers and accessibility. ACM, pp 52–59. http://doi.org/10.1145/1090785.1090797

38. May E (1961) Music for deaf children. Music Educ J 47(3):39–42. http://doi.org/10.2307/3389102

39. McCarthy Music's Illuminating Piano (2015) McCarthy music's illuminating piano: learn to play the piano with McCarthy music. https://www.mccarthypiano.com. Accessed 13 Sep 2015

40. McCord K, Fitzgerald M (2006) Children with disabilities playing musical instruments. Music Educ J 46–52

41. Mitroo JB, Herman N, Badler NI (1979) Movies from music: visualizing musical compositions. In: Proceedings of the 6th annual conference on computer graphics and interactive techniques. ACM, pp 218–225. http://doi.org/10.1145/800249.807447

42. Mori J, Fels DI (2009) Seeing the music can animated lyrics provide access to the emotional content in music for people who are deaf or hard of hearing? In: Science and technology for humanity (TIC-STH), 2009 IEEE Toronto international conference, pp 951–956. http://doi.org/10.1109/TIC-STH.2009.5444362

43. Music animation machine (2016) Music worth watching. http://www.musanim.com/. Accessed 19 March 2016

44. Music and the deaf (2015) Enriching lives through music. http://matd.org.uk/. Accessed 1 May 2015

45. Myles K, Binseel MS (2007) The tactile modality: a review of tactile sensitivity and human tactile interfaces. DTIC Document

46. Nanayakkara S, Taylor E, Wyse L, Ong SH (2009) An enhanced musical experience for the deaf: design and evaluation of a music display and a haptic chair. In: Proceedings of the

SIGCHI conference on human factors in computing systems. ACM, pp 337–346. http://doi.org/10.1145/1518701.1518756

47. Novich SD, Eagleman DM (2015) Using space and time to encode vibrotactile information: toward an estimate of the skin's achievable throughput. Exp Brain Res 1–12. http://doi.org/10.1007/s00221-015-4346-1

48. Palmer R (1997). Feeling the music philosophy: a new approach in understanding how people with sensory impairment perceive and interpret music. http://www.russpalmer.com/feeling.html

49. Palmer R (2015) Music floors. http://www.russpalmer.com/music.html. Accessed 10 May 2015

50. Palmer R (2016) Tac-tile sound system. http://www.russpalmer.com/tactile.html. Accessed 20 March 2016

51. Petry B, Illandara T, Nanayakkara S (2016) MuSS-bits:sensor-display blocks for deaf people to explore musical sounds. In proceedings of the 28th Australian conference on computer-human interaction (OzCHI '16), 72–80. https://doi.org/10.1145/3010915.3010939

52. Pouris M, Fels DI (2012) Creating an entertaining and informative music visualization. In: Miesenberger K, Karshmer A, Penaz P, Zagler W (eds) Computers helping people with special needs. Springer, Berlin, pp 451–458. http://doi.org/10.1007/978-3-642-31522-0_68. Accessed 26 Oct 2015

53. Rashid R, Aitken J, Fels DI (2006) Expressing emotions using animated text captions. In: Miesenberger K, Klaus J, Zagler WL, Karshmer AI (eds) Computers helping people with special needs. Springer, Berlin, pp 24–31. http://doi.org/10.1007/11788713_5. Accessed 15 March 2016

54. Reed CM, Doherty MJ, Braida LD, Durlach NI (1982) Analytic study of the tadoma method further experiments with inexperienced observers. J Speech Lang Hear Res 25(2):216–223

55. Roebuck J (2007) I am a deaf opera singer. The Guardian. http://www.theguardian.com/theguardian/2007/sep/29/weekend7.weekend2. Accessed 26 Aug 2015

56. Schlaug G (2015) Musicians and music making as a model for the study of brain plasticity. In: Finger S, Boller F, Altenmüller E (eds) Progress in brain research. Elsevier, pp 37–55. http://www.sciencedirect.com/science/article/pii/S0079612314000211. Accessed 29 July 2015

57. Shibata D (2001) Brains of deaf people "hear" music. Int Arts-Med Assoc Newslett 16:4

58. Shinohara K, Tenenberg J (2009) A blind person's interactions with technology. Commun ACM 52(8):58–66. http://doi.org/10.1145/1536616.1536636

59. Stumpf C (1883) Tonpsychologie. Leipzig: Hirzel 1

60. Tan HZ, Robert Gray J, Young J, Traylor Ryan (2003) A haptic back display for attentional and directional cueing. Haptics-e 3(1):1–20

61. Verillo RT (1991) Vibration sensing in humans. Music Percept 9(3):281–302

62. Vy QV, Mori JA, Fourney DW, Fels DI (2008) EnACT: a software tool for creating animated text captions. In: Miesenberger K, Klaus J, Zagler W, Karshmer A (eds) Computers helping people with special needs. Springer, Berlin, pp 609–616. http://doi.org/10.1007/978-3-540-70540-6_87. Accessed 15 March 2016

63. WHO (2015) Deafness and hearing loss. WHO. http://www.who.int/mediacentre/factsheets/fs300/en/. Accessed 6 Aug 2015

64. Wyse L, Nanayakkara S, Seekings P, Ong SH, Taylor E (2012) Perception of vibrotactile stimuli above 1 kHz by the hearingimpaired. In: NIME'12

65. Yang H-J, Lay Y-L, Liou Y-C, Tsao W-Y, Lin C-K (2007) Development and evaluation of computer-aided music-learning system for the hearing impaired. J Comput Assist Learn 23(6):466–476. http://doi.org/10.1111/j.1365-2729.2007.00229.x

66. Zatorre RJ, Chen JL, Penhune VB (2007) When the brain plays music: auditory–motor interactions in music perception and production. Nat Rev Neurosci 8(7):547–558. http://doi.org/10.1038/nrn2152

67. Zhou Y, Sim KC, Tan P, Wang Y (2012) MOGAT: mobile games with auditory training for children with cochlear implants. In: Proceedings of the 20th ACM international conference on multimedia. ACM, pp 429–438. http://doi.org/10.1145/2393347.2393409

Teach Me How! Interactive Assembly Instructions Using Demonstration and In-Situ Projection

Markus Funk, Lars Lischke, Sven Mayer, Alireza Sahami Shirazi
and Albrecht Schmidt

1 Introduction

In industrial settings production effectiveness and efficiency is paramount. Over the last 50 years automation and robotics massively changed how goods are manufactured. It is foreseen that over the next decade a further revolutionary shift to more flexible production systems will happen, as outlined in the smart factory [35] and Industry 4.0 [27] initiatives. However in most domains production is not fully automated and human workers still play an essential role. For example in the car industry, human workers are cooperating with robots in complex assembly processes. With individualized products many variants are produced in the same production line at the same time. Also, as storage costs are increasing, ordered products are produced on demand—just when they were ordered. This process is called lean manufacturing. However, in such flexible production environments where many different variants of a product are assembled, the task of the worker becomes more and more complex. Humans are creative and have great skill for manipulating objects. However dealing with large number of variants is cognitively demanding and typically high level instructions are required (*this screw should be attached to this part*). Low level instructions (e.g. how to hold the screw, how to insert it into a hole, how to hold the screw driver, etc.) are not required as these motor-cognitive tasks are simple for humans (in contrast to a robot) [48]. Workers have to understand which variant they are creating and what steps are required. With small lot sizes and frequent changes, classical training and teaching approaches do not scale. Neither learning all possible variants upfront, nor getting a traditional training session each time the product on the assembly line changes is a viable option. The method of choice is to provide the information required for the production when the worker needs them.

The majority of the work has been conducted while he was a researcher at the University of Stuttgart.

M. Funk (✉) · L. Lischke · S. Mayer · A. Schmidt
University of Stuttgart, Pfaffenwaldring 5a, 70569 Stuttgart, Germany
e-mail: makufunk@hotmail.com

A.S. Shirazi
Yahoo Inc., 701 1st Ave, Sunnyvale, CA 94089, USA

© Springer Nature Singapore Pte Ltd. 2018
J. Huber et al. (eds.), *Assistive Augmentation*, Cognitive Science
and Technology, https://doi.org/10.1007/978-981-10-6404-3_4

In traditional production with large lot sizes and a small number of variants it was useful and cost-effective to create training and information material upfront. Depending on the task and environment assembly manuals were created as paper-based instructions or videos. More recently assembly instructions were also created for in-situ systems, e.g. Pick-by-Light [21] or Augmented Reality (AR) systems [3, 8, 42]. The cost for creating instructions can be divided by the number of products created based on this instruction. Consider the following example of assembling a refurbished starter for a car (similar to the one used in the study). The average assembly time for the product by a worker is about 3 min. If we assume the creation of a traditional tutorial video, this will take 120 min, creating written instructions takes 60 min, and for a set of instruction based on demonstration we estimate 6 min. For a lot size of 10.000 the cost of creating the instruction is less of an issue as the assembly time will be the major cost factor. However for a lot size of 20 it is clear that the creation of instructions becomes a major issue. Skill acquisition for individuals and skill transfer within the workforce becomes more important and a major factor for competitiveness in flexible production environments. In our research, we envision that skills of workers can be captured with little or no effort and can be transferred to others to pick them up with little effort. Continuing the example from above, we assume that for the starter with the lot size of 20 a skilled worker would do one assembly to remind herself of the best way of doing it, then she would assemble a second one where the system is used to record the assembly, and then the remaining 18 starters could be assembled by untrained workers. In this chapter, we empirically compare two approaches for recording and using of the instructional material: videos and interactive assembly instructions, which are semantically rich and where the information is embedded and presented step by step.

Extending our previous system [16, 17], we have developed a functional system that automatically generates these interactive assembly instructions using the Programming by Demonstration (PbD) approach (Fig. 1). While the user demonstrates

Fig. 1 The system provides visual instructions for supporting workers during the assembly of an engine starter. It highlights the position where a part should be assembled and checks if it is assembled correctly. Instructions were created using a simple programming by demonstration approach

an assembly task by assembling the parts step by step, our system detects the currently obtained part and the position where it is assembled. Using this information, the system automatically creates an assembly instruction, where the semantics of each step is retained. With this PbD approach, interactive in-situ instructions can be created almost as fast as recording a video of assembling the product but still retain all features of interactive instructions. Our approach enables the instructor to physically show a new workflow to the system, same as it would be shown in front of a camera or to a new worker. Further, the system can use the recorded information to provide step-wise instructions using in-situ projection. It highlights the bin where a part should be picked from and the position where it should be assembled in each step. Therefore, our system provides a new means for process engineers for creating interactive instructions and a new way for workers to use assembly instructions. We believe that this work adds to the area of assistive augmentation by introducing a stationary assistive system that provides cognitive assistance during assembly tasks.

The contribution of this chapter is threefold: first, we present a system that automatically detects work steps and creates a semantically rich assembly instruction while an assembly is performed using a depth camera. The system also uses in-situ projection to provide the steps for assembling a product. Second, we compare video based instructions and Augmented Reality-based step-by-step in-situ projection. With 32 participants using reproducible tasks of different complexity, we compare the impact of the different representations for the quality and performance of the assembly. The study shows that, especially for more complex assemblies, the error rate (ER) decreases, the assembly is faster, and the mental demand is reduced using in-situ instructions. Third, we investigate the effort required for creating instructions in a realistic work environment with industrial workers. We present the findings of a user study with expert users comparing three approaches for creating assembly instructions: traditional video recording, using a graphical editor, and automated extraction of instruction using the described systems. The results show that using our system, assembly instructions can be created faster with less perceived cognitive load in comparison to using a graphical editor while the effort is comparable to traditional video recording. We validate the created instructions in an industrial setting with 51 workers in a car production plant.

The chapter is structured as follows: after reviewing the prior work, we present an interactive assembly system that creates semantically rich assembly instructions from a demonstration. We describe a laboratory study in which we compare video based instruction and the in-situ projection. Then two studies in an industrial setting, one for creating instructions and one for using instructions, are reported. The participants in the studies are skilled workers for creating the instructions and unskilled workers for using the created instructions. As a task, we use the assembly of a refurbished car starter. Finally, we finish the chapter with discussing implications.

2   Related Work

Creating and providing assembly instructions using interactive systems has been the subject of various research. In the following, we provide an overview in relevant research areas for creating and presenting interactive assembly instructions, namely, Programming by Demonstration, projected surfaces, and Augmented Reality.

2.1   Programming by Demonstration

PbD (also referred to as programming by example) was initially proposed to enable users to record macros without knowing any programming language or writing code. This approach has been adopted by many application domains which comprise desktop applications like MS Excel, computer-aided design, and text editing [33]. Thereby, a user's actions are translated into a textual procedure, which later can be played back and altered. For example, the Peridot system [37] enables interface designers to demonstrate how a UI should look like rather than having to program it. Recently, Kubitza and Schmidt [31] introduced a framework that enables non-programmers to use PbD to program for smart environments.

The PbD approach is also used to teach new motion sequences to humanoid robots by recording movements of a human worker. Aleotti et al. [1] reproduce and optimize measured trajectories of a human worker. The trajectories can then be used to infer high level actions [6]. After defining actions, the sequence of the actions can be played back and altered. Instead of programming physical robots, Marinos et al. [36] use a PbD approach to rapidly create animations for a virtual robot inside a blue or green box of a virtual studio.

Overall, previous work shows that even non-programmers can use a PbD approach for creating digital content, programming physical robots, and defining procedures. This rapid creation of digital content does not need special training as the actions that are performed by the users are natural actions that users would also do without using a computer. In our system, combining the PbD approach with interactive surfaces and AR, we enable users to create interactive projected instructions for humans.

2.2   In-Situ Projection and Interactive Surfaces

Projecting information directly into the interaction space or onto objects has been used to augment real world objects with digital information or to display information in-situ. Pinhanez [38] uses a rotating mirror to create displays out of arbitrary surfaces and to augment objects with information. Combining this technology with a camera, projected surfaces become interactive. In the Touchlight system [44], Wilson uses two RGB-cameras and computer vision techniques to detect touch input on a projected surface. The LuminAR [34] system integrates such a camera-projector system into an anglepoise lamp. The lamp can project information next to a recognized object on a desk. Furthermore, it can detect performed gestures. On the other hand, other projects applied in-situ projection to different areas, e.g. the kitchen [32] or sterile training areas [39].

With the proliferation of depth cameras, sensing interaction on projected surfaces has become easier. Wilson [45] suggests an algorithm that enables sensing of multi-touch without using an RGB image. This algorithm was improved and provided as a framework in the Ubi Displays toolkit [24]. With this toolkit a user can define multiple touch-enabled areas that have their own projected content. The dSensingNI project [28] combines Wilson's algorithm with gestural user inputs in their tabletop system. Furthermore, they support detecting the presence, volume and orientation of cubical objects using a top-mounted Kinect. Although dSensingNI is capable of detecting stacked objects, the system is not able to detect if a construction is correctly assembled.

Overall, related work showed how to augment physical objects with digital information using in-situ projection. Further, user interaction on these projected displays can be detected using RGB or depth cameras. Our system also uses top projection to provide in-situ information. Additionally, it can detect if a construction is assembled correctly using a depth camera and computer vision.

2.3 Providing Assembly Instructions for Training Workers

Videotaping of a manual assembly process is a straightforward approach for creating assembly instructions, which is used to teach assembly procedures to untrained workers. These so-called Utility Videos (e.g. Memex[1]) are produced by professional companies for training unskilled or new workers in a new assembly task. On the other hand, systems providing interactive AR instructions [11] have been suggested to assist workers during assembly tasks. For example, Pick-by-Light systems visually show the worker, where the next part has to be picked from [3], or how a part has to be assembled [4]. Also Head-Mounted Displays (HMDs) can show the worker the next part and where the part has to be assembled [10, 23, 42]. More recently, assistance technologies focused on projecting instructions directly into the workers field of view (e.g. Light Guide Systems[2]). This in-situ projection reduces the complexity of the given feedback, as it is projected directly into the work space, instead of giving feedback on an external monitor. Such projected instructions are usually created using a graphical editor. However, with frequently changing variants of the same product, creating and maintaining instructions is cumbersome. Instead of being able to just alter the changed steps of the variant's workflow, the instructor often needs to change the whole workflow as even small changes effect succeeding work steps.

---

[1]http://memex-academy.eu/ (last access 03-18-2016).

[2]http://www.ops-solutions.com/ (last access 03-18-2016).

2.4 Augmented Reality in Assembly

Industrial Augmented Reality (IAR) is now almost always present in a manufactured product's life-cycle. Experiencing a designed product can be done immediately [13], industrial robots cooperating with human workers can be programmed using AR-debugging approaches [12], order picking can be supported using HMDs [22] or projector carts [21], and maintaining existing machines and products can be supported directly on site [46]. Workers can even be motivated during the work tasks by using IAR for gamification [30].

Prior work has used AR to provide assembly instructions. An overview about this topic is presented by Büttner et al. [9]. A strand of work has augmented parts of a product with sensors. Antifakos et al. [2] use instrumented tools and assembly parts to infer a user's current action and suggest proactive instructions for assembling an IKEA PAX wardrobe. Compared to a printed manual, their system can dynamically react upon a user's action as it is aware of all possible assembly orders rather than printing one fixed order. However, integrated sensors may influence the design of the product.

Instead of augmenting the assembly parts, other research proposed mobile systems for displaying interactive assembly instructions by augmenting the users with sensors. For example, Ward et al. [43] equip the user with body worn microphones and accelerometers to infer the user's current activity in an assembly environment. Even when combining multiple features [7] to recognize an activity more reliably, a body worn system unfortunately cannot detect if a part is assembled correctly.

Using HMDs is another approach that has been explored to display assembly instructions during work tasks [11]. It has been shown that it can reduce the task completion time (TCT) and mental workload [42]. This concept has been adapted to several domains. Through a user study, Henderson et al. [26] report that users have less head movements using HMD-based AR instructions while repairing a vehicle. Zauner et al. [47] use AR markers to provide assembly instructions on a HMD for assembling furniture. Salonen et al. [40] are also using a marker approach while experimenting control modalities. However, overall the feedback on these assembly instruction systems has to be explicitly advanced to the next work step.

While the aforementioned approaches are for mobile settings, assisting systems for stationary setups have been explored, too. For example, Bannat et al. [3] present a framework using a top-mounted RGB camera to detect bins automatically based on their color and shape. Once the position of the bins in known, their system uses the RGB camera to detect the position of the worker's hand. In their system assembly instructions are shown on a monitor close to the work area. The system highlights the next bins to pick from using a top-mounted projector. Korn et al. [29] extends this approach by using a top-mounted depth camera instead of a RGB camera and a top-mounted projector in production environments. The position of the bins and the position of an assembled part have to be defined manually using a graphical editor. Their system then highlights the bin to pick from. As their system cannot automatically detect the correct assembly in each step, it uses projected buttons that the user can manually advance the projection to next step. Recently, Funk et al. [16, 17] investigated the potentials of using in-situ projected instructions at the workplace to support workers with cognitive impairments. They found that using in-situ projected instructions workers with cognitive impairments can assemble more complex products without increasing the time or errors per work step. Further, they found that using a simple contour-based highlighting as assembly instruction is perceived as better than video, pictorial, and no instructions [19].

Overall, previous work suggests using a setup consisting of a top-mounted projector and a depth camera to display instructions using in-situ projection. In our system, we also use that setup to detect picking from bins. We additionally use the depth camera to detect if the assembly is performed correctly. In contrast to prior work, our system automatically creates in-situ feedback based on a demonstrated assembly and automatically highlights the bin to pick from. Overall, by using PbD, our system requires no additional effort from the user when creating instructions except assembling the product.

3 Instructions Creation Through Demonstration

We developed an interactive system for creating and providing semantically-rich assembly instructions, which uses the concept of PbD to create instructions. Hereby, the system is able to automatically create instructions for an assembly task while it is being performed. It detects out of which bin a part is picked and where the part is assembled. During the assembly, the system can project assembly instructions directly into the work area. Accordingly, it highlights which bin to pick a part from, and at which place it should be assembled on the workpiece carrier. In the following, we give an overview about hardware and software of the system, which is an extension of the software presented in [16].

3.1 Hardware Setup

We designed our system that it reflects an assembly workplace found in the industry. Figure 2 shows the system and its components. It consists of a top-mounted projector, a Kinect depth sensor, a number of bins, and a workpiece carrier. The bins in the back of the system (Fig. 2c) contain the assembly parts. Tools needed for the assembly task are placed at the side of the system (Fig. 2e). The steel plate in Fig. 2d is a workpiece carrier that holds parts during the assembly. In an industry setup, workpiece carriers are exchanged between work places using a skate wheel conveyor and then are fixed from below using a pneumatic clasp. We firmly mounted the workpiece carrier on the table to prevent it from moving while conducting work steps.

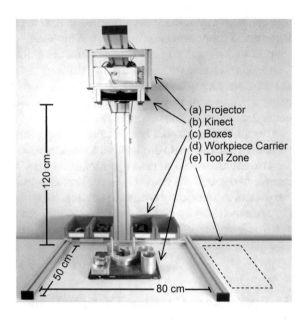

Fig. 2 Our system consists of four components: **a** top-mounted LED-projector, **b** kinect for windows, **c** bins containing the assembly parts, and **d** the workpiece carrier holding the product. The area in **e** depicts the tool zone

To make our system transportable, we built an aluminum frame which holds the Kinect, the projector, and the bins. The projector highlights the bin from which the user should pick a part from, tools that should be used, and the position on the work-piece carrier where the part has to be assembled. The Kinect detects if a part was picked from a bin, if a part is assembled correctly, and if a tool is used. The number of bins, the content of the bins, and the workpiece carrier change according to the manufactured product and the steps that are performed at the work place. With our current setup, the system can handle a maximum of eight bins (2 rows × 4 bins) due to the limited angle of the Kinect which has to cover both work area and the bins. Our system provides a predefined layout of the bins and a predefined area for putting tools on the right side of the system. The layout of the bins and the area for putting tools can be changed and customized using a graphical interface.

In our setup, we use an Acer K335 LED-projector with 1000 ANSI Lumen and a Kinect for Windows running on a depth resolution of 320 × 240 pixels with 30 frames per second. Both Kinect and projector are mounted 120 cm above the surface and are facing the table. They are calibrated using the 4-point calibration of the Ubi Displays toolkit [24].

3.2 Work Step Detection

For detecting assembly steps and creating instructions, we define a high-level rep-resentation of performed actions (c.f. [6]). We call this high-level representation a

(a)   (b)   (c)

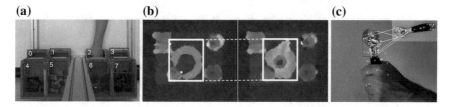

Fig. 3 Overview about the triggers used in our concept: **a** detecting a hand entering a bin, **b** detect object placement based on depth data, **c** recognizing the presence and the absence of an object compared to a previously taken reference image using computer vision

workflow, which consists of a finite number of work steps. Each work step has an initial state and a trigger condition (trigger) for advancing to the next step. A trigger is activated by one of the following three actions: (1) pick a part from a bin, (2) assemble a part that was just picked, (3) use a tool on the product.

Using the trigger concept, actions in each step of the assembly can be detected. Further it enables implicit interaction [41] with the system to trigger the next step of the workflow. In our model, we define three triggers which notify the system that one of the actions was performed (see Fig. 3).

3.2.1 Pick Detection

Our system uses the top-mounted Kinect to detect when a hand enters a bin. The placement of the bins can be defined using a graphical interface where the user can adjust the size and the position of the bin directly in the Kinect's RGB image (Fig. 3a). Once the position of the bins is defined, the system stores a depth map of the bin's area to continuously compare it to the most recent depth image. Using this technique, we can define 3D cubes in the work area that layover the bins. When the system registered a pick from a bin, the bin is briefly highlighted by the projector. To be robust against depth sensor noise, we consider at least 4 mm changes in depth value. Our algorithm compares each depth pixel inside the cube to the previously stored state. If the participant picks a part from a bin, the percentage of changed depth pixels becomes larger. If the percentage exceeds a threshold, the bin is triggered. The threshold is dependent on the size of the worker's hand, the size of the bin, and the distance of the camera to the bins. In an informal test we found that a threshold of 63% is a good value to reliably detect the hands of 5 different persons.

3.2.2 Assembly Detection

At the beginning of recording a work step, the system captures the initial depth data as an initial state. Then, the worker can start assembling the product. To capture each work step correctly while recording an instruction, the worker needs to step out of the work area after each assembly step. The system's built-in movement detection converts color frames into a gray-scale image and subtracts each 15th frame from the previous one. If a difference between the images was found, the system knows that the worker is still performing a task. If no movement was detected for the last 1.5 s, the system assumes that the worker's hands are out of the work area and captures the current depth data. This data is compared to the previously captured initial state by transforming both depth arrays into gray-scale images and subtracting them from each other using EmguCV.[3] This algorithm enables the system to detect where a part was assembled (see Fig. 3) and to distinguish between removing and adding parts. If the area changed is larger than a threshold of 150 pixels in total, it is considered to be a valid work step. The threshold of 150 pixels was chosen empirically and provided a robust trade-off between filtering sensor noise and detecting assembly steps. Afterwards, the latest depth data is stored with the work step as a desired state and visual feedback is given to the user.

When playing back a workflow, the depth data from the desired state is continuously compared to the depth data of the current frame by comparing each pixel. If the current depth frame matches the desired state, the work step is considered to be performed correctly and the system proceeds to the next work step.

3.2.3 Detection of Tool Usage

In our prototype, we predefined a tool zone at the right side of the system, which can be changed and customized. Our system continuously scans the depth data of the defined tool zone and checks the changes in the data. In case, the change in depth data is over a threshold of 63%, our algorithm runs the SURF object recognition algorithm [5] to compare the image of the defined zone to the previously recorded reference image of the object (Fig. 3c). If the object cannot be recognized in the picture, it is considered to be taken and the system assumes that the worker is using it. When the user puts the object back at its place, the depth data changes and the system runs the SURF algorithm again. If the object is recognized again, the system triggers that the object was used and displays visual feedback.

3.2.4 Resulting Instructions

As the system is detecting the object that is picked, the assembly which was performed, and the tool which was used, a semantic description of the performed step is stored for each step. In particular the information about which part is picked and which tool is used helps to add flexibility. Using the resulting instructions can be transferred to another production table (or to multiple tables) where there may be different arrangement for parts and tools. Changing the location of a bin with parts (either automatically detected or manually entered into the system) can then be used

---

[3]http://www.emgu.com/ (last access 03-18-2016).

to change an existing instruction. Assume a bin containing screws is moved from left to right. As the semantic information is available the visual feedback showing the worker where to pick can be moved to the new position accordingly. In contrast, video-based instructions would need to be re-recorded.

3.3 In-Situ Projection for Visual Feedback

In our system, visual feedback for the next work step is created automatically from the user's actions. As the Kinect and the projector are calibrated, the system knows which pixel in the Kinect's image matches which pixel of the projector. For picking objects out of a bin, the system highlights the correct bin with a green light. Once the user picked a part from the bin, the system projects green light at the position where the part has to be put on the workpiece carrier. Thereby, the highlighted position is calculated automatically by comparing the depth data of the initial and the desired state. As suggested by previous work [19], the system calculates a contour visualization. When a part should be removed from the workpiece carrier, the system highlights the contour of the part on the workpiece carrier with a red light. In case a tool should be used, the system highlights the tool's position with a yellow light in the tool zone.

3.4 Playing Back Workflow Instructions

The triggers are also used for playing back a previously recorded workflow. The system plays each step of a workflow in the order it was defined. If the current step is a picking-task from a bin, the system only advances to the next step if the defined bin is triggered. In case the current task is assembling a part, the system advances when the depth data stored for the desired state matches with the current depth data to an extend of at least 90% accuracy. When the current step is to use a tool, the system checks if the tool is removed from its place. If the tool was absent for at least three seconds and it is put back again, the system considers the object as used and triggers the next step.

4 Study #1: Assembly with Different Complexities

To assess our system using assembly tasks with a different number of steps and complexity, we conducted a user study in our laboratory. Inspired by previous work [15, 16, 18, 42], we decided to use Duplo[4] bricks for creating construction models with

---

[4]http://www.lego.com/en-us/duplo (last access 03-18-2016).

(a)   **(b)**   **(c)**   **(d)**

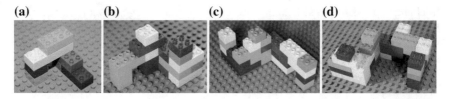

Fig. 4 The constructions used in the lab study with four different complexity levels: **a** 8 bricks, **b** 16 bricks, **c** 24 bricks, and **d** 32 bricks

different numbers of bricks. As the system can monitor a maximum of eight bins, we considered four models with four different numbers of bricks, i.e., 8, 16, 24, and 32. All the four models were created using 8 different types of bricks in five different colors. They all have one arch in the bottom level. Figure 4 shows the four constructions.

4.1 Method

A mixed design was considered for carrying out this study. We used a between-subject design with the type of instruction as the only independent variable with two levels: the video-based approach and the in-situ projection approach. Within the groups, we used a repeated measures design with the number of bricks as independent variable (4 levels). As dependent variables in both groups, we measured the ER, the TCT, and the NASA-Task Load Index (NASA-TLX) score. The order of the repeated measures tasks was counterbalanced according to the Balanced Latin Square.

We created two assembly instructions for each construction model: recording the video, and using the PbD approach. For recording the video instructions, we used a camcorder and videotaped the assembly instructions in HD resolution recorded from over the shoulder of the worker. For recording the projected instructions, we used our PbD system. In both cases one of the researchers performed the assembly task while the instructions were recorded and created. For both conditions, the content of the bins and the bins' arrangement were identical. Each type of brick had a separate bin resulting in 8 different bins.

For the video instruction, a monitor was placed next to the work area (see Fig. 5a). The participant could play and pause the video using the space key on the keyboard at any time during the assembly. For the in-situ projection, the participant sat in the same place in front of the plate and for each step instructions were projected into the work area by either highlighting a bin to pick from or the position a brick should be placed (see Fig. 5b).

The procedure of the study was as follows: after welcoming the participant and giving a brief introduction about assembling products, we collected the demographics. Then, one of the instructions was assigned to the participant. When the participant

(a)   (b)

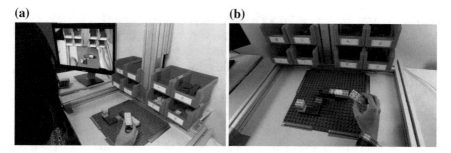

Fig. 5 The setup of the two conditions used in the lab study. The video condition uses a monitor to display the instructions (**a**). The PbD condition projects visual feedback onto the Duplo bricks (**b**)

was ready, the experimenter started the instruction and measured the TCT. The participant was instructed to only use the predominant hand to pick and assemble the bricks as assembling two parts at the same time is not supported by the system. The whole experiment session including hands and the picking from the bins was video recorded for each participant. After assembling each model, the participant was asked to fill in the NASA-TLX questionnaire. The participant repeated this procedure for all four construction models. After the study, two researchers independently watched the videos and counted the errors for each participant. They compared the results and in case of inconsistency, the researchers reviewed the videos together until they came to an agreement.

We recruited 32 participants, 8 female and 24 male with the average age of 25.1 years ($SD = 3.9$) using the University's mailing list. All participants were students in various majors. They had no prior knowledge in assembling the Duplo buildings nor participated in the two previous studies. Furthermore, none of the participants was colorblind. The study was conducted in our lab at the University of Stuttgart.

4.2 Results

We statistically compared the ER, the TCT, and the NASA-TLX score between the four models and the two instruction methods conducting a two-way mixed ANOVA. Mauchly's test indicated that the assumption of sphericity had been violated for ER ($\chi^2(5) = 17.60, p < 0.004$) and TCT ($\chi^2(5) = 23.29, p < 0.001$). Therefore, the degrees of freedom were corrected using Greenhouse-Geisser estimated of sphericity ($\epsilon = 0.73$ for ER and $\epsilon = 0.68$ for TCT). The t-test with Bonferroni correction was considered as post hoc test for all cases.

The analysis revealed that the difference in the ER between the four models was not significant ($F(2.18, 65.36) = 1.94, p > 0.05$). The model with the 24 steps had the largest ER ($M = 0.66$, SD 1.61) followed by the 32-step model ($M = 0.59$, $SD = 1.38$) and 16-step model ($M = 0.47$, $SD = 1.04$). Whereas, the effect on the ER between the two feedback approaches was statistically significant ($F(1, 30) = 11.20$, $p < 0.002, r = 0.39$). The effect size estimated shows a medium and hence substantial effect. The post hoc test showed that the video-based instruction had a significantly larger ER than the in-situ projection instruction ($M = 0.86$, $SD = 1.36$ vs. $M = 0.05$, $SD = 1.36$, $p < 0.002$).

Analyzing the TCT between the constructions showed that it statistically significantly differed ($F(2.05, 61.5) = 217.88$, $p < 0.001$). Post hoc tests revealed a significant difference between all constructions. The 32-step model had the longest TCT ($M = 2.31$ min, $SD = 0.69$) followed by 24-step ($M = 1.83$ min, $SD = 0.70$) and 16-step ($M = 1.10$ min, $SD = 0.31$). Such differences were already expected due to the variation in the number of bricks. On the other hand, feedback approaches had statistically significant effect on the TCT ($F(1, 30) = 63.82$, $p < 0.001$, $r = 0.80$). The effect size indicates a large and substantial effect. Surprisingly, the TCT using the video method took 1.5 times longer than the PbD method ($M = 1.73$ min, $SD = 0.45$ vs. $M = 1.08$ min, $SD = 0.45$).

Furthermore, there was a statistical significant difference in the NASA-TLX score between the constructions ($F(3, 90) = 3.63$, $p < 0.01$). The post hoc tests showed that the difference was only significant between the 8-step and 32-step models ($M = 22.34$, $SD = 16.20$ vs. $M = 27.87$, $SD = 17$, $p < 0.1$). The score between other constructions was not significant (all $p > 0.05$). The average score for the 16-step model was 25.03 ($SD = 17.47$) and for the 24-step construction the score was 26.38 ($SD = 16.93$). The comparison between the methods revealed a statistical significant effect on the mental load ($F(1, 30) = 19.73$, $p < 0.001$, $r = 0.54$). The effect size indicates the effect is large and substantial. The mental load for the in-situ instruction approach was 60% smaller than the video-based instruction ($M = 15.62$, $SD = 25.96$ vs. $M = 35.19$, $SD = 25.96$, respectively).

4.2.1 Impacts of Number of Steps in Assembly

We further assessed the differences between the two feedback approaches for different complexities, i.e. having a different number of assembly steps. To achieve this, for each construction model, we conducted the t-test between the video and in-situ instructions and pair-wise compared the ER, TCT, and NASA-TLX score. The Levene's test conducted in all cases to test the equality of variances. In case the assumption was violated the degrees of freedom were adjusted.

The comparison of ER showed the in-situ instruction had the fewer errors than the video instruction in all levels of complexity (see Fig. 6a). The difference was not significant in the 8-step ($t(15) = 1.86$, $p > 0.05$, $r = 0.43$) and 16-step constructions ($t(16.84) = 1.94$, $p > 0.05$, $r = 0.42$). But, the difference was statistically significant in the 24-step construction ($t(15) = 2.48$, $p < 0.05$, $r = 0.53$) and the 32-step construction ($t(15) = 2.31$, $p < 0.05$, $r = 0.50$). The effect size estimate indicates that the effect on ER for all four models using the provided instructions is large.

The comparison of TCTs revealed that the difference between both approaches was significant for all steps except for the 8-step construction ($t(30) = 1.11$, $p > 0.05$,

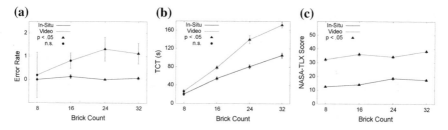

Fig. 6 The results of the lab study for constructions with different number of steps: **a** error rate (ER), **b** task completion time (TCT), and **c** NASA-Task Load Index (NASA-TLX) score

$r = 0.20$). Figure 6b shows the average TCT for the four constructions using the two instruction methods. In all cases the TCT was significantly faster using the in-situ approach (for the 16-step montage: $t(30) = 4.69$, $p < 0.001$, $r = 0.65$; for 24-step montage: $t(30) = 5.67$, $p < 0.001$, $r = 0.71$; for 32-step montage: $t(30) = 7.92$, $p < 0.001$, $r = 0.68$). The effect sizes show that effect of the provided instructions on the TCT of the assembly tasks is substantial except for the 8-step assembly task.

Further, the NASA-TLX scores statistically significantly differ in all four constructions (see Fig. 6c). In all cases the score for the in-situ instruction was significantly lower than the video approach: for the 8-step montage, $t(19.37) = 4.30$, $p < 0.001$, $r = 0.70$; for 16-step montage, $t(20.52) = 4.58$, $p < 0.001$, $r = 0.71$; for 24-step montage, $t(30) = 2.90$, $p < 0.007$, $r = 0.47$; for 32-step montage: $t(30) = 4.37$, $p < 0.001$, $r = 0.62$. The effect size estimate indicates that the effect on the perceived cognitive load using the two instruction approaches is large, and therefore substantial for all models.

4.3 Discussion

The results of the analysis reveal that there are significant differences between the video instruction and the in-situ projection approach in the ER, the TCT, and the perceived cognitive load during the assembly tasks. Using the in-situ system, the ER decreases up to 17%, the TCT is up to 1.5 times faster, and the perceived cognitive load is reduced up to 60% in comparison to the video-based instruction.

Further, the comparison of the in-situ and video-based instructions in different levels of complexity unveil that the in-situ instruction outperforms the video-based approach independent of the number of steps. In all levels the ER is lower and the TCT is faster. These differences are significant when the number of steps in the assembly task increase. On the other hand, the perceived cognitive load is significantly lower for the in-situ instruction independent from the number of steps in the assembly task.

5   Study #2: Creating Assembly Instructions

To evaluate our system for creating assembly instructions in a real world scenario, we conducted a user study using a real assembly task (a refurbished car's engine starter) with industrial workers. We made this conscious choice to increase the validity of the results, even if it is harder to reproduce the results. Using students and a lab-based study is in our view not appropriate in to address this questions.

5.1   Method

We used a repeated measures design with three conditions for creating an instruction: by demonstration using our system, using the editor, and video recording. The only independent variable was the creating-method. As dependent variables, we measured the task completion time (TCT) for creating instructions and the NASA-TLX score [25]. The order of the conditions was counterbalanced.

For the editor condition, we re-implemented the system presented by Korn et al. [29]. In contrast to our system, the user should use a graphical user interface (GUI) to manually highlight the bins, the workpiece carrier, or tools that have to be used for the assembly task using different geometric shapes (see Fig. 7a). Further, the GUI is used to define actions in each step of the assembly and create an instruction. For the video condition, we recorded a video of the assembly from the worker's point of view. A camcorder was installed behind the user in such a way that the worker's point of view could be simulated. The participant had to inform the experimenter when the video recording should be started and stopped.

As the assembly task, we chose the assembly of a car's engine starter (see Fig. 1). The task consisted of five steps and in each step one part should be assembled. When all five parts were put together on the workpiece carrier, the worker should fix two screws on top of the starter using a screwdriver.

We carried out the study in a car manufacturing company in Germany. After welcoming the participant and explaining the course of the study, we collected the demographics. Next, we introduced the participant to the workpiece carrier and let them get familiar with it. We allowed the participant to assemble the engine starter twice to get themselves familiar before starting the study. Afterwards, the study was started and participants had to create instructions using the three approaches. End of each condition, the experimenter measured the TCT. Afterwards, the participant completed the NASA-TLX questionnaire. At the end, we collected qualitative feedback through semi-structured interviews.

We recruited 10 workers from the company (2 female, 8 male), who were familiar with the engine starter. The participants were aged between 17 and 53 years ($M = 32.1$, $SD = 13.9$). All participants had experience in assembling the engine starter for at least one year and could be considered as experts.

(a)   **(b)**

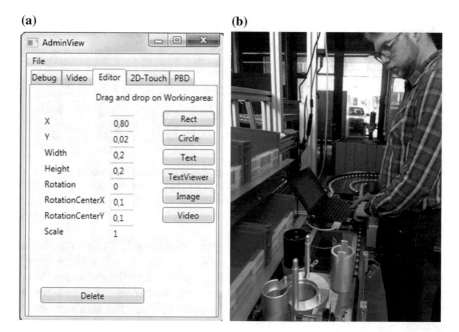

Fig. 7 **a** The graphical editor allows changing the properties of projected elements. **b** The worker can adjust the projection directly on the workpiece carrier

5.2   Results

We statistically compared the TCT between the methods. Mauchly's test indicated that the assumption of sphericity had been violated ($\chi^2(2) = 18.04$, $p < 0.0001$). Therefore, the degree of freedom was corrected using Greenhouse-Geisser estimates of sphericity ($\epsilon = 0.52$). A repeated measures ANOVA showed that there is a statistically significant difference in TCT between the methods ($F(1.05, 9.49) = 256.04$, $p < 0.0001$, $r = 0.97$). The effect size estimate reveals a large and therefore substantial effect. Post hoc tests using Bonforroni correction revealed a significant difference between all three methods (all $p < 0.05$). The video method had the shortest TCT ($M = 0.58$ min, $SD = 0.08$) followed by the PbD ($M = 1.52$ min, $SD = 0.63$) and the editor ($M = 16.16$ min, $SD = 3.07$). The results are also depicted in Fig. 8a.

Further, we statistically compared the NASA-TLX scores between the methods (see Fig. 8b). The sphericity assumption was not violated ($p > 0.05$). A repeated measures ANOVA determined that the methods used had a statistically significant effect on the NASA-TLX score ($F(2, 18) = 19.83$, $p < 0.0001$, $r = 0.81$). The effect size estimate shows a large and substantial effect. Post hoc tests using the Bonferroni correction revealed that the editor had a statistically significantly higher perceived cognitive load ($M = 23.10$, $SD = 7.79$, $p < 0.007$) than PbD ($M = 11.80$, $SD = 5.22$) and the video ($M = 10.40$, $SD = 7.07$, $p < 0.001$). However, the difference between PbD and video was not statistically significant ($p > 0.05$).

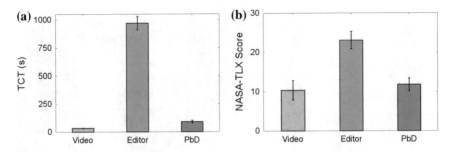

Fig. 8 The results of the user study for creating assembly instructions **a** The task completion time across the different approaches. **b** The perceived cognitive load across the approaches using the NASA-TLX. The *error bars* depict the standard error

The qualitative feedback showed that the participants found the editor hard to use. Although they were experts in assembling an engine starter, they didn't have enough experience in using a computer (e.g., P6, P1). Further, a participant stated "*using the editor is too time-consuming*" (P3). One participant had also privacy concerns when recording a video as co-workers could identify him based on his hands and his wristwatch (P4).

5.3 Discussion

The results of the study reveal that the editor approach requires significantly higher perceived cognitive load compared to the PbD and video approaches. Whereas, there is no significant difference in perceived cognitive load required for creating assembly instructions using the PbD and video approaches. Hence, the additional perceived cognitive load added due to the use of our interactive system is not significant.

On the other hand, the results show that recording the video is the fastest way for creating an assembly instruction followed by PbD and the editor approach. One reason is that no additional time is required to capture the depth information after each assembly step. In contrast, the PbD approach requires that the users shortly remove their arms and head from the work area to capture the depth data of the product.

Although the PbD-based and video-based approaches are faster and require less cognitive effort than the editor-based approach in creating instructions, the approaches might differ when assembling the engine starter. Therefore, we conducted a followup study to evaluate the instructions while assembling the engine starter with novice users.

6 Study #3: Evaluation of Assembly Instructions

In the previous study we assessed different approaches for creating assembly instructions. In order to evaluate the practicality of the approaches in assembling a product, we conducted a followup study assembling the same engine starter we used in the previous study using the previously created instructions.

6.1 Method

For providing assembly instructions we used the instructions created in the previous study. We randomly chose one instruction created using each approach resulting in three instructions: (1) the video-based assembly instruction, (2) the in-situ projection instruction created using the editor, (3) the in-situ projection instruction created using PbD. For the in-situ projection instruction using the editor, the user explicitly created the instruction using a graphical editor. In contrast, our system automatically generated the other instruction. We chose a between subject design with three groups to prevent a learning effect between the different instructions. The only independent variable that differed between the groups was the type of instruction. As dependent variables we measured the number of errors (ER), the task completion time (TCT), and the NASA-TLX score.

We conducted the study in the same company as in the previous study. After welcoming the participant and explaining the course of the study, we collected the demographics and ensured that the participant never assembled an engine starter before. Then, the participant was accompanied to our prototype and one of the instructions was assigned and explained. As the participants did not differ in skills, the condition was randomly assigned. The participant was told to assemble an engine starter based on the instructions provided. When the participant was ready, the experimenter started the instruction and counted the ER. The TCT was measured by the system automatically. During the assembly the experimenter did not provide any help. After the assembly was done, the participant was asked to fill in a NASA-TLX questionnaire. Finally, qualitative feedback was collected through a semi-structured interview.

We recruited 51 participants (12 female, 39 male) aged between 23 and 60 years ($M = 47.8$, $SD = 9.3$). We divided the participants equally between the conditions, resulting in 17 participants per condition. All participants were employees of the company and were unfamiliar with the assembly task and the product, i.e., assembling an engine starter. Hence, they can be considered novice users. None of the recruited participants took part in the previous study.

6.2   Results and Discussion

We statistically compared the ER, the TCT and NASA-TLX between the groups. The assumption of homogeneity of variance had not been violated ($p > 0.05$). A one-way ANOVA test revealed no significant effect on ER between the groups ($F(2, 48) = 0.89, p > 0.05$). The group using the instruction created by our system had the lowest ER ($M = 1.12$, $SD = 0.86$) followed by the group using the instruction created by the editor ($M = 1.24$, $SD = 0.90$) and the group using the video-based instruction ($M = 1.53$, $SD = 1.01$). Results are also depicted in Fig. 9b

The statistical analysis also revealed no significant difference in the TCT between the groups ($F(2, 48) = 0.32$, $p > 0.05$). According to Fig. 9a, the group using the instruction created by our system had the shortest TCT ($M = 2.21$ min, $SD = 1.05$) and the group using the instruction created with the editor had the longest TCT ($M = 2.52$ min, $SD = 1.39$). The group using the video instruction took on average 2.22 min ($SD = 1.31$) to assemble the product.

The analysis showed no significant effect on the NASA-TLX score between the groups ($F(2, 48) = 1.38, p > 0.05$). The group using the video-based instruction had the lowest perceived cognitive load ($M = 20.59$, $SD = 13.90$) followed by the group using the instruction created using our system ($M = 27.53$, $SD = 13.87$) and using the editor ($M = 28$, $SD = 15.84$). A graphical representation is depicted in Fig. 9c.

The qualitative feedback indicated that the projected instructions were generally well perceived. They particularly found the step by step feedback of the projected instructions very helpful (P42, P33). Additionally they mentioned that directly projected feedback onto the workplace was very useful (P30, P12). One participant stated that "I don't have to think anymore while working" (P24). Another participant mentioned that "I would rather work autonomous in the daily life, but for training I would use it" (P45). Participants using the video instruction mentioned that the video was helpful for learning the task instead of having an instructor (P22, P51) but they didn't want a video playing all day (P37, P25).

The analysis shows that the ER is reduced and the TCT is faster in the assembly task using our system compared to the other two approaches. However, the change is not significant. The results indicate that the instruction automatically created using

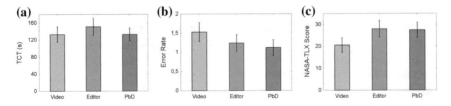

Fig. 9 The results of the user study for evaluating the previously created assembly instructions **a** the task completion time across the different approaches. **b** The error rate for the different approaches. **c** The perceived cognitive load across the approaches using the NASA-TLX. The *error bars* depict the standard error

our system slightly performs better than the explicitly created instruction using the editor. On the other hand, the results suggest that the in-situ projection increases the perceived cognitive load required during the assembly, but the difference is not significant. The qualitative feedback indicates that the step by step instructions provided directly in the work area through the in-situ projection is more accepted than a video-based instruction.

As the assembly instruction only consisted of five steps, no big differences were expected. Based on the results from study 1 it can be expected that with more steps the differences between these instructions would increase and a clearer advantage for the in-situ instructions would be expected to show.

7 Implications

The aforementioned user studies revealed implications on both creating assembly instructions and performing an assembly task based on previously created instructions. In the following we discuss the insights gained through these studies.

7.1 Creating Assembly Instructions

The results of the study indicate that creating instructions using the editor approach is more time-consuming and demands more cognitive load compared to the PbD approach and the video-based approach.

Using the PbD approach, the time required to create an instruction is higher than the video recording approach as our system requires the user to wait 1.5 s between each step for detecting that a step was performed. The time is even higher when using the editor approach as the user has to manually specify each step. However, editing steps in both PbD approach and editor approach is easier than editing video-based instructions since each step can be modified separately. In contrast, video-based instructions need to be post-processed and manually edited. Editing videos can be complex and may result in re-recording the video even if only a single step needs to be altered. Another advantage of both in-situ approaches is that it records the depth information of each step. Using the depth information the system can monitor if the correct part is picked and if it is assembled correctly.

While the video approach has the lowest perceived cognitive load when creating instructions, the results further show that using the PbD approach does not significantly increase the mental load. However, the editor approach induces a higher perceived cognitive load by interacting with GUIs.

A further advantage of including semantic information into the instructions is that the instructions can be targeted to a specific work place automatically. One could even imagine a skilled worker in one company (or one country) can create the instructions and these instructions can then be downloaded to an assembly table in another company (or country) (cf. [17]).

7.2 Assembly Performance

When it comes to assembling a product, the results suggest that the in-situ projection approach reduces the mental load of the worker, the TCT, and the ER. Specially, these effects are significant when the number of assembly steps increase. As the in-situ assembly instructions are provided directly in the work area, the distraction is minimized compared to showing the videos on a monitor close to the work area. This reduces the cognitive effort that is required for following instructions and also reduces the TCT for assembling a product. Furthermore, our system's step by step error control can monitor if the correct part is picked and if it is assembled correctly using depth information. This leads to fewer errors even when the number of steps increases.

7.3 Limitations

It should be mentioned that the proposed PbD system has certain limitations. The current version of the system tracks only one assembly part per work step. This process is favored by the industry as it is less error prone than assembling multiple parts in a single step. However, the system can be easily extended to track more than just a single item per step. Furthermore, all assembled parts on the workpiece carrier should be visible to the top-mounted Kinect to monitor the assembly task. Therefore, the workpiece carrier has to be designed to support this setup.

8 Conclusion

In this chapter, we presented a system that leverages the concept of PbD to create semantically-rich assembly instruction for enabling assistive augmentation at the workplace. The proposed system enables process engineers who are creating assembly instructions to create instructions faster than using a graphical editor for creating assembly instructions. In contrast to just recording video, which is slightly faster, our proposed system retains all features of interactive instructions and does not add any significant perceived cognitive load to the worker using the instructions for learning assembly steps in comparison to watching video instructions. The system was evaluated with experts in a production environment using a real product.

The system provides instructions using in-situ projection directly in the work area. It highlights a bin where a part should be picked from and shows the position where the part should be assembled. In a large laboratory study, we could show that in-situ instructions outperform the video-based instruction in assembly tasks with different numbers of steps. It decreases the error rate, the task completion time, and the perceived cognitive load. This was also validated in a real assembly environment.

Creating such interactive instructions based on demonstration is not only limited to the assembly work place. It could be easily ported to other application domains. We currently explore further domains, in particular the home environment, where we assess if such a system could teach persons with learning disabilities to learn basic skills for independent living, such as cooking [14, 32] and cleaning their home.

Acknowledgements This work is funded by the German Federal Ministry for Economic Affairs and Energy in the project motionEAP [20], grant no. 01MT12021E. We thank Mathias Hoppe for his work in helping to implement the software and conducting the user study. We further thank Klaus Klein, Michael Spreng, and Johann Hegel from Audi AG.

References

1. Aleotti J, Caselli S (2006) Robust trajectory learning and approximation for robot programming by demonstration. Robot Auton Syst 54(5):409–413
2. Antifakos S, Michahelles F, Schiele B (2002) Proactive instructions for furniture assembly. In: UbiComp 2002: ubiquitous computing. Springer, pp 351–360
3. Bannat A, Gast J, Rigoll G, Wallhoff F (2008) Event analysis and interpretation of human activity for augmented reality-based assistant systems. In: 4th international conference on intelligent computer communication and processing, 2008. ICCP 2008. IEEE, pp 1–8
4. Barna J, NovakovaMarcincinova L, Novak-Marcincin J, Fecova V, Janak M, Torok J (2012) Open source tools in assembling process enriched with elements of augmented reality. In: Proceedings of the 2012 virtual reality international conference. ACM, p 2
5. Bay H, Tuytelaars T, Van Gool L (2006) Surf: speeded up robust features. In: Computer vision-ECCV 2006. Springer, pp 404–417
6. Billard A, Calinon S, Dillmann R, Schaal S (2008) Robot programming by demonstration. In: Springer handbook of robotics. Springer, pp 1371–1394
7. Blanke U, Schiele B, Kreil M, Lukowicz P, Sick B, Gruber T (2010) All for one or one for all? Combining heterogeneous features for activity spotting. In: 2010 8th IEEE international conference on Pervasive computing and communications workshops (PERCOM Workshops). IEEE, pp 18–24
8. Büttner S, Sand O, Röcker C (2015) Extending the design space in industrial manufacturing through mobile projection. In: Proceedings of the 17th international conference on human-computer interaction with mobile devices and services adjunct. ACM, pp 1130–1133
9. Büttner S, Mucha H, Funk M, Kosch T, Aehnelt M, Robert S, Röcker C (2017) The design space of augmented and virtual reality applications for assistive environments in manufacturing: a visual approach. In: Proceedings of the 10th international conference on pervasive technologies related to assistive environments. ACM, pp 433–440
10. Büttner S, Funk M, Sand P, Röcker C (2016) Using head-mounted displays and in-situ projection for assistive systems: a comparison. In: Proceedings of the 9th ACM international conference on pervasive technologies related to assistive environments. ACM, p 44
11. Caudell TP, Mizell DW (1992) Augmented reality: an application of heads-up display technology to manual manufacturing processes. In: Proceedings of the twenty-fifth hawaii international conference on system sciences, vol 2. IEEE, pp 659–669

12. Collett T, MacDonald BA (2006) Developer oriented visualisation of a robot program. In: Proceedings of the 1st ACM SIGCHI/SIGART conference on human-robot interaction. ACM, pp 49–56

13. Fiorentino M, de Amicis R, Monno G, Stork A (2002) Spacedesign: a mixed reality workspace for aesthetic industrial design. In: Proceedings of the 1st international symposium on mixed and augmented reality. IEEE Computer Society, p 86

14. Funk M, Korn O, Schmidt A (2015) Enabling end users to program for smart environments. In: Proceedings of the CHI 2015—workshop on end user development in the internet of things era 12.2, pp 9–14

15. Funk M, Kosch T, Schmidt A (2016) Interactive worker assistance: comparing the effects of in-situ projection, head-mounted displays, tablet, and paper instructions. In: Proceedings of the 2016 ACM international joint conference on pervasive and ubiquitous computing. ACM, pp 934–939

16. Funk M, Mayer S, Schmidt A (2015) Using in-situ projection to support cognitively impaired workers at the workplace. In: Proceedings of the 17th international ACM SIGACCESS conference on computers and accessibility

17. Funk M, Schmidt A (2015) Cognitive assistance in the workplace. Pervasive Comput IEEE 14(3):53–55

18. Funk M, Kosch T, Greenwald SW, Schmidt A (2015) A benchmark for interactive augmented reality instructions for assembly tasks. In: Proceedings of the 14th international conference on mobile and ubiquitous multimedia. ACM, pp 253–257

19. Funk M, Bächler A, Bächler L, Korn O, Krieger C, Heidenreich T, Schmidt A (2015) Comparing projected in-situ feedback at the manual assembly workplace with impaired workers. In: Proceedings of the 8th ACM international conference on pervasive technologies related to assistive environments. ACM, p 1

20. Funk M, Kosch T, Kettner R, Korn O, Schmidt A (2016) Motioneap: an overview of 4 years of combining industrial assembly with augmented reality for industry 4.0. In: Proceedings of the 16th international conference on knowledge technologies and datadriven business

21. Funk M, Shirazi AS, Mayer S, Lischke L, Schmidt A (2015) Pick from here!: an interactive mobile cart using in-situ projection for order picking. In: Proceedings of the 2015 ACM international joint conference on pervasive and ubiquitous computing. ACM, pp 601–609

22. Guo A, Raghu S, Xie X, Ismail S, Luo X, Simoneau J, Gilliland S, Baumann H, Southern C, Starner T (2014) A comparison of order picking assisted by head-up display (HUD), cart-mounted display (CMD), light, and paper pick list. In: Proceedings of the 2014 ACM international symposium on wearable computers. ACM, pp 71–78

23. Hahn J, Ludwig B, Wolff C (2015) Augmented reality-based training of the PCB assembly process. In: Proceedings of the 14th international conference on mobile and ubiquitous multimedia. ACM, pp 395–399

24. Hardy J, Alexander J (2012) Toolkit support for interactive projected displays. In: Proceedings of the 11th international conference on mobile and ubiquitous multimedia. ACM, p 42

25. Hart SG, Staveland LE (1988) Development of NASA-TLX (task load index): results of empirical and theoretical research. Adv Psychol 52:139–183

26. Henderson S, Feiner S (2011) Exploring the benefits of augmented reality documentation for maintenance and repair. IEEE Trans Visual Comput Graph 17(10):1355–1368

27. Hermann M, Pentek T, Otto B. Design principles for industrie 4.0 scenarios: a literature review

28. Klompmaker F, Nebe K, Fast A (2012) dSensingNI: a framework for advanced tangible interaction using a depth camera. In: Proceedings of the sixth international conference on tangible, embedded and embodied interaction. ACM, pp 217–224

29. Korn O, Schmidt A, Hörz T (2013) Augmented manufacturing: a study with impaired persons on assistive systems using in-situ projection. In: Proceedings of the 6th international conference on pervasive technologies related to assistive environments. ACM, p 21

30. Korn O, Funk M, Abele S, Hörz T, Schmidt A (2014) Context-aware assistive systems at the workplace: analyzing the effects of projection and gamification. In: Proceedings of the 7th international conference on pervasive technologies related to assistive environments. ACM, p 8

31. Kubitza T, Schmidt A (2015) Towards a toolkit for the rapid creation of smart environments. In: End-user development. Springer, pp 230–235
32. Lee C-H, Bonnani L, Selker T (2005) Augmented reality kitchen: enhancing human sensibility in domestic life. In: ACM SIGGRAPH 2005 posters. ACM, p 60
33. Lieberman H (2001) Your wish is my command: programming by example. Morgan Kaufmann
34. Linder N, Maes P (2010) LuminAR: portable robotic augmented reality interface design and prototype. In: Adjunct proceedings of the 23nd annual ACM symposium on user interface software and technology. ACM, pp 395–396
35. Lucke D, Constantinescu C, Westkämper E (2008) Smart factory-a step towards the next generation of manufacturing. In: Manufacturing systems and technologies for the new frontier. Springer, pp 115–118
36. Marinos D, Wöldecke B, Geiger C (2013) Prototyping natural interactions in virtual studio environments by demonstration: combining spatial mapping with gesture following. In: Proceedings of the virtual reality international conference: laval virtual. ACM, p 2
37. Myers BA (1986) Creating dynamic interaction techniques by demonstration. In: ACM SIGCHI bulletin 17.SI, pp 271–278 (1986)
38. Pinhanez C (2001) The everywhere displays projector: a device to create ubiquitous graphical interfaces. In: Ubicomp 2001: ubiquitous computing. Springer, pp 315–331
39. Rüther S, Hermann T, Mracek M, Kopp S, Steil J (2013) An assistance system for guiding workers in central sterilization supply departments. In: Proceedings of the 6th international conference on pervasive technologies related to assistive environments. ACM, p 3
40. Salonen T, Sääski J, Hakkarainen M, Kannetis T, Perakakis M, Siltanen S, Potamianos A, Korkalo O, Woodward C (2007) Demonstration of assembly work using augmented reality. In: Proceedings of the 6th ACM international conference on image and video retrieval. ACM, pp 120–123
41. Schmidt A (2000) Implicit human computer interaction through context. Pers Technol 4(2–3):191–199
42. Tang A, Owen C, Biocca F, Mou W (2003) Comparative effectiveness of augmented reality in object assembly. In: Proceedings of the SIGCHI conference on human factors in computing systems. ACM, pp 73–80
43. Ward JA, Lukowicz P, Troster G, Starner TE (2006) Activity recognition of assembly tasks using body-worn microphones and accelerometers. IEEE Trans Pattern Anal Mach Intell 28(10):1553–1567
44. Wilson AD (2004) TouchLight: an imaging touch screen and display for gesture-based interaction. In: Proceedings of the 6th international conference on Multimodal interfaces. ACM, pp 69–76
45. Wilson AD (2010) Using a depth camera as a touch sensor. In: ACM international conference on interactive tabletops and surfaces. ACM, pp 69–72
46. Wohlgemuth W, Triebfürst G (2000) ARVIKA: augmented reality for development, production and service. In: Proceedings of DARE 2000 on designing augmented reality environments. ACM, pp 151–152
47. Zauner J, Haller M, Brandl A, Hartman W (2003) Authoring of a mixed reality assembly instructor for hierarchical structures. In: The second IEEE and ACM international symposium on mixed and augmented reality, 2003. Proceedings. IEEE, pp 237–246
48. Zollner R, Rogalla O, Dillmann R, Zollner M (2002) Understanding users intention: programming fine manipulation tasks by demonstration. In: IEEE/RSJ international conference on intelligent robots and systems, 2002, vol 2. IEEE, pp 1114–1119

Augmented Narrative: Assisting the Reader with Sound

Susana Sanchez Perez, Naohito Okude and Kai Kunze

1 Introduction

Designing an assistive augmentation for narrative texts requires a solution-based creative process that begins with a deep understanding of the topic at hand. Therefore, it is essential to study literature and how narrative texts are perceived to answer questions as to why readers need assistance and how to implement the assistive augmentation.

The study and analysis of literature is a scholarship that corresponds to literary theory, with its different schools that discuss and write about writing. One of these schools is the reader-response criticism that views literature as a performing art where readers create their own text-related performance. This division of literary theory began in the 1960s, and considers the meaning of a narrative text is completed through interpretation [3]. Amongst the different views of reader-response theorists, Individualists focus only on the reader's experience, while Uniformists assume that text and reader have a shared responsibility to convey meaning. This makes reading both, subjective and objective. One must look into reading processes to create meaning, and experience to understand the narrative text. Here, there are two levels of understanding: the information explicitly presented in the narrative text, and the integration of the different pieces of information from that narrative text.

The reader-response criticism is based on Kintsch's construction-integration model [37]. This general theory contends that comprehension arises from the interaction between the narrative texts to comprehend, and the reader's general knowledge and lived experiences. In fact, studies based on the construction-integration model stress the importance of background knowledge and the reader's ability to generate

S. Sanchez Perez (✉) · N. Okude (✉) · K. Kunze (✉)
KMD, Keio University, Tokyo, Japan
e-mail: su_saps@hotmail.com

N. Okude
e-mail: naohito.okude@gmail.com

K. Kunze
e-mail: kai.kunze@gmail.com

© Springer Nature Singapore Pte Ltd. 2018
J. Huber et al. (eds.), *Assistive Augmentation*, Cognitive Science
and Technology, https://doi.org/10.1007/978-981-10-6404-3_5

inferences to fill in gaps of missing information, supporting the idea of a multidi-mensional scale of text complexity [12, 23].

The construction-integration model has many similarities with the possible world (PW) theory, adapted to narratives texts by Lewis [42]. Lewis postulates the idea of "modal realism" to make a distinction between real worlds and the actual world. Here, all PW are real, as they exist even if it is only in the imagination, but there is only one actual world (AW). This AW serves as a model to mentally construct other storyworlds that can differ from the AW. Readers imagine fictional worlds as close as possible to the AW, changing only what the narrative text mandates, in an interpretative rule called "the principle of minimal departure". For example, if the narrative text describes Pegasus as a winged stallion, the reader will image a creature that resembles in every respect a real world horse, real world meaning the AW, except for the fact that this horse has wings [61].

Contrary to the reader-response criticism are the text-oriented schools. Formalism claims readers can understand narrative texts while remaining ambivalent about their own culture. However, if said readers belong to a culture where horses do not exist, because of their geographical location, would they be able to understand Pegasus? In this case, the lack of previous knowledge and experience could affect the perception of the narrative text, frustrating readers that do not have enough information to men-tally construct the mythological creature. Another example is the Victorian novel, which represents a culture and social norms of old fashion didacticism [39]. Set in a historical present, period literature turns out to be problematic to today's readers, since this literary genre confers them with a simple lesson from the past that has become meaningless with time [67]. The perception of the Victorian novels is then affected by the culture of contemporary readers, who would need assistance to fill in the gaps of information about context and social norms of Victorian societies in order to extract meaning.

That being the case, "Augmented Narrative" is proposed as an assistive augmenta-tion. In line with the reader-response criticism this augmented narrative also uses the construction-integration model as framework. The augmentation allows the narrative text to provide an embodied experience to the reader, supplementing information for the lack of previous lived experiences, and thus assisting the multidimensional scale of the narrative text complexity.

2 Embodiment

Augmented Narrative looks at embodiment to assist perception and retrieve meaning from the narrative text, using the qualities of sound to add non-verbal information to the narration. The sonic augmentation allows an embodied experience, where the reader perceives the narrative through mind and body: the written word and sound.

The phenomenologist Alfred Schtz calls literary meaning "monothetic", since it relies on idealizable and objectifiable semantic content that makes the written word time transcendent, sign-oriented, and conceptual [63]. But to truly conceive reading as a performing art, we have to look for a "polythetic" meaning. To illustrate, theatrical performances are a form of communication, and more appropriately of storytelling, which do not rely on semantic content. In theatre, meaning is time-immanent, fully perceived in an embodied experience of the moment [17]. In this way, plays transmit a 'polythetic' meaning, assisting audiences to experience within the embodied consciousness, and not as an object of consciousness.

Narratological studies agree that orality and sound, which are in a close connection with theatre and the natural world, have nothing in common with literature. This field of literature considers the experience of the world is detached from the omniscience of the written word, as literature offers readers the artificial effect of experiencing and viewing outside the natural world [16]. By doing so, narratology has failed as a comprehensive discipline as regardless of literature's disconnection with its oral past, it still shares some characteristics that allow for an embodied experience.

Embodiment, in a simple definition, is the biological and physical presence of the body [43]. However, Maurice Merleau-Ponty, phenomenological philosopher, defines embodiment by separating the objective body, the physiological entity, from the phenomenal body, the sense of ones motor capacities with which to experience and perceive the world [49]. Merleau-Ponty's definition of embodiment has served to connect phenomenology to cognitive sciences and neuroscience, necessary fields of study to understand how readers retrieve meaning from written language. The connection between mind and body also concerns literary theory, as the brain is able to simulate actions through language processing. Read words automatically elicit neural activations similar to those that occur during the perception of events, because verbs evoke mental representation of objects to which the action refers [45].

In parallel with the construction-integration model, neuroscientists have found evidence that the brain has two dimensions: mirroring and self-projection. Mirroring allows to physically resonate what others are experiencing, whereas self-projection implies imagining what should be felt, and then attributing the imagined experiences to others. Imaginative capacity, which involves previous knowledge for self-projection, is required to attenuate the cognitive distance gap, as mirroring is only felt in low intensity [36]. Mirroring allows an intuitive and immediate comprehension of actions, and self-projection uses an inferential process of self experience to reason intentions into emotions [9]. In Augmented Narrative sound is used to assist imagery to further attenuate the cognitive gap between mirroring and self-projection, augmenting the experience of the phenomenal body to physically resonate the storyworld.

The embodiment postulation is a remediation of print to convey a polythetic meaning through a co-evolutionary process in a double dimension of the interface: human-media and media-media [64]. The proposed embodied reading experience follows Andy Clark's "extended mind" theory, where cognition is a cycle that runs from brain through body, world, and back. In fact, Clark sees the boundary of "skin and skull" as cognitively meaningless. The human-media interface is a *biofeedback*, a text-body interaction where the narrative text knows when to assist the reader by measuring attention and memory work. On the other hand, the remediation of the media-media interface follows a coevolution where different media has contaminated one another to create the *sonic assistance*[48]. Theatrical sound design is combined with the latest available technology to reinterpret literature as an embodied experience. Here, sound operates in a cognitive ecology of multidimensional content, where the reader uses mind and body to read, listen, feel, remember, and imagine.

3   Cognitive Ecology of Literature

The changes made to transform literature into a medium that assists its reader affects not only the object, in this case the book, but also the social structure around it and the cognitive architecture of written communications. Therefore, it is important to consider how the object's assistive augmentation will affect the processes of how literature is created and consumed, as no technology can serve as a reading space in the absence of writers and readers. The framework of cognitive ecology is proposed because it supports extending the medium's capability to be an assistive storyteller within the simplest communication model of sender-receiver. Furthermore, the framework is used to translate literary theory into cognitive and computational terms bringing literature into the field of HCI.

Edwin Hutchins defines cognitive ecology as the study of context, in particular the mutual dependence between elements in an ecosystem [29]. Here, the mind arises within a physical system distributed over space and time. Hutchins suggests sensory and motor processes are not peripheral, making the relations of brain-and-body interactions with the environment an important unit of analysis. This ecological approach uses the distributed cognition theory to describe, in computational terms, human-work-systems where knowledge lies not only within the individual, but the individuals social and physical environment [27]. The goal of the theory is to describe how distributed units are coordinated by analyzing the interactions between individuals, the media used, and the environment within which the activity takes place.

The focus of a cognitive ecology of literature is in the sender-receiver model, by augmenting the medium's capability to better transmit the message in absence of the writer, who is not in the same physical space as the reader. When the reader has not enough background knowledge to model the narrative world, the result is a lack of understanding that hinders the ability of the writer to transmit her message; if there is not enough shared understanding between the two, communication cannot take place. This brings antagonism to the reader, drifting attention away from the narrative text, making it difficult to continue reading.

For the writer, her readers are unknown, leaving it up to these readers to retrieve meaning through decoding and interpretation. The medium of print allows the narrative to talk to the reader, but not to listen. The reader can interpret, but will not be reassured or assisted. However, if the medium changes into an empathetic narrator capable to understand the reading experience, then literature becomes an intrinsic co-constitution in a dialog between three agents: author, reader, and book. In Hutchins' terms, the unit of writer affects the reader through the narrative text, whereas the unit of reader affects the writer on how her work is interpreted. Here, there is a shared responsibility to achieve meaning, even in print, but with different weights. The difference is that in print both units, writer and reader, have a delimiting line that suppresses, not only the natural dialog found in oral communications, but also relegates auditory and any other sensuous complexity. For that reason, it is important to look into Plato's advices to see the boundaries between units more as joints, where connectivity is no longer relatively long, but strong and reciprocal. The three agents: writer, reader, and book, can then interact within a cognitive ecology of literature where perception and attention are technologically mediated, in a system distributed across social structures, objects, and cognitive architectures. The assistive remediation comes when literature is augmented to become aware of the human cognitive architecture: sensory memory, working memory and long-term memory through a biofeedback.

The cognitive ecology principle has been used to create a framework for Shakespearean Studies [69]. It served to analyze theatre across a system of neural and psychological mechanisms, bodily and gestural norms, physical environments, cognitive artefacts, and technologies of sound and light; elements that affect and modulate each other. In this theatrical cognitive ecology sound design is used to mediate the audience's attention and perception during the play. Here, sonic information is detected as perceivable opportunities for action in the environment, where sound psychoacoustics allow audiences to distinguish between a variety of sounds, and how to interact with them [55].

In this cognitive ecology sound is used to lessen the communication problem between writer and reader increasing the available information about the storyworld. The ecological approach is based on perception of information rather than sensation [18]. Even though ecological psychology seems to contradict information processing (recognition is a multi-stage process between perceptual qualities of sound source, abstract representation in memory, meanings, and associations), it explains why sound can carry meaning even if that same sound has been created artificially, as it is concerned with the invariant properties of the sound [19, 66]. For instance, one is able to recognize someone's voice on the phone, even if that person has a terrible cold.

4 Biofeedback

Based on the relation between mind and body, computers are capable to understand emotional states. Consequently, computers can take actions by recognizing likes and dislikes of users, evolving from mere tools to personal companions. Sensors give logic and reason to mechanical devices enabling them to empathize with their users, looking for positive and negative reactions to a task. These sensors have been used to measure the user's emotional stand in digital entertainment to allow interaction in narratives. For instance, Interactive Storytelling captures user's emotions to create an affect-based interactive narrative. This affective gaming creates new experiences by adapting the game to the player's emotions in two modalities: managing the story-line to achieve the game's emotional goal or adapting the narrative only to generate positive emotions [74].

Rather than adapting the narrative text to balance emotions, an augmented narra-tive focuses on keeping the reader engaged without the need of multiple storylines or changing the author's work. Here, sensors are used to measure the reader's cognitive responses, reflected in the body, to what is being read. As the proposed augmen-tation is concerned with assisting cognitive processes of perception and attention rather than emotional responses, the biofeedback is based on mental workload. Men-tal workload increases with memory work, used in reading comprehension, where previous knowledge is required, and correlates to engagement tasks of sustained attention [4]. Thus, the biofeedback looks for levels of engagement to know when to assist. Based on the reader's mental workload, the augmented narrative can detect if the reader has been or not engaged in the storyworld.

A higher mental workload signals the reader is being engaged in the story. On the other hand, a low mental workload signals a lack of engagement, when the reader is finding difficulties to create her own mental performance of the narrative text, and is in need of assistance. Consequently, the proposed biofeedback operates as a bridge for understanding between author, medium, and reader, re-designing the medium to be an assistive narrator that acts and delivers the story according to the readers needs. In this way, authors can be assured that their work will be meaningful to their readers.

Perception and emotion are intricately related, and the line between the two is blurred at the best of times. For instance, even though engagement is related to atten-tion and memory work, is a positive feeling. While there are different physiological metrics to create a biofeedback, some might be more adequate to find emotional states and others to find cognitive processes. To design a biofeedback for an aug-mented narrative, the physiological metrics of mental workload in the heart rate vari-ability (HRV) and nose temperature are examined. The first of these metrics served to design a circumflex model of engagement, while the second was used to develop smart-glasses to detect reading engagement.

4.1 Metrics of Engagement

The study of the empathetic relations readers build with fictional characters has focused in verbal feedback. For example, measuring engagement with expressions such as: "*I felt I was there right with Phineas*" about the book "A Separate Peace" by John Knowles [58]. The verbal feedback is possible since books, with their authorial omniscience place non-natural characters into a natural frame, allowing authors to engage readers with the artificial effect of experiencing and viewing outside the nat-ural world. In turn, readers attribute a mental stance to these non-natural characters, in the same way they do in everyday life, building mental reconstructions of read emotions and actions [2, 21].

Empathy is the mind-reading ability that allocates mental states to fictional characters and is essential for engagement [26]. The segment of the story affects the degree of engagement, which is more prominent during the climax of the story, where the reader is expected to be paying more attention [1]. This deeper sense of involvement is part of Csikszentmihalyi's flow theory [11]. Here, positive experiences come when engaged in task demands where the person is in deep sense of control, and the activity feels rewarding. Flow can also be considered to involve straining tension and mental workload. However, the stress found in flow is a positive experience referred to as eustress. Based in this theories an in order to understand the demanding character of flow activities, and find engagement in reading tasks, some studies in the metrics of the heart are reviewed.

4.1.1 Heart Rate Variability

Involuntary responses controlled by parasympathetic components slow the heart rate (HR) and sympathetic components raise it. In the absence of arousal, attraction and aversion can be detected in the variations between heartbeats in the heart rate variability (HRV), called valence [35]. Engagement can be associated with a decrease in HR, contrary to the increment found in emotional responses. Thus, information of engagement can be determined by a positive valence, detected in the ratio of low frequency energy to high frequency energy that represents the extent of sympathetic and parasympathetic influence in the HRV. On the other hand the highest frequency indicates boredom, in negative valence, reflecting the lowest mental workload.

Mental workload is affected by engagement, as information-processing increments mental workload. Keller et al. looked into the impact of flow in conditions of boredom, fit, and overload while completing knowledge tasks with questions from the TV show "Who wants to be a millionaire?". Results revealed the highest HRV was found during boredom, reflecting the lowest mental load, while a decreased HRV reflected involvement in the fit condition, showing that higher levels of flow can be associated with a low HRV. In the absence of arousal, the valence of different conditions of involvement such as engagement, boredom, and overflow, can be detected in the variations between heartbeats. In the HRV, the variations between heartbeats, time interval is referred as the QRS complex in which the ratio of low frequency energy to high frequency energy represents the extent of sympathetic and parasympathetic influence on the heart.

To find the metrics of engagement, Keller's frequency data was compared to McCraty's et al. typology of six HRV patterns that denote different modes of psychophysiological interaction [47]. The graphic from McCraty et al. describes wave forms of emotional states, divided into normal every day life emotional experiences and hyper-states of emotional experiences. These were compared to Keller et al. HRV data using the axis of arousal and balance. In the findings 'fit' corresponds to serenity, while 'boredom' corresponds to apathy. Even though both scholars look for different processes, cognitive and emotional, the wave patterns are similar, linking emotional and cognitive metrics of the heart.

Fig. 1 Circumflex model of
engagement based on the
metrics of HRV

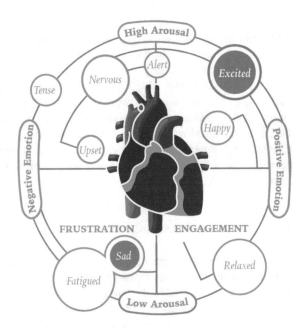

Based on this findings, a circumflex model of engagement is suggested (See Fig. 1). In this model, starting clockwise, quadrant four and quadrant one represent arousal, an emotional stimulations by the narrative text. In quadrant two the reader is in a state of relaxation or eustress, meaning engagement. Finally, in quadrant three, the system has an input of frustration or stress, which in this case will be defined as non-engagement. Quadrant two and three are low arousal, meaning there is an absence of demanding tasks in emotional processing, where the mental workload takes over, and is regarded as a cognitive process involving attention.

4.1.2 Nose Temperature

The metrics of facial temperature have been found to reliably discriminate between positive and negative emotions, and cognitive tasks. Here, the sympathetic system is responsible of lowering the temperature of the nose in mental workload tasks. On the other hand, frustration, which is related to boredom, increases the blood volume into supraorbital vessels along with the skin temperature of the nose [25, 31].

Israel Waynbaum, in his vascular theory of emotional expression, attributes the experience of emotions to follow facial expressions rather than preceding them; relating it to the James-Lange theory [7, 73]. The vascular theory is based on the fact that the supply of blood to the brain and face comes from the same source, the carotid artery. Therefore, the reactions to circulatory perturbations in the facial artery produce disequilibrium in the cerebral blood flow. The facial muscles contract and push against the skull bone structure, acting as a tourniquet on arteries and veins.

This serves to regulate the blood flow, affecting the cerebral blood flow, reducing or complementing it. Blood flow alterations cause temperature changes that modify the neurochemistry of the brain. The thermoregulatory action influences peptides and neurotransmitters, temperature dependent, found to produce emotional changes. Cooling is associated to pleasant and warming to unpleasant feelings [8, 13, 30]. For example, facial skin temperature of nose, forehead, and cheeks, decreases when laughing and increases when being angry. The decrease in nose temperature can be the most dramatic, dropping as much as 2.0 °C in 2 min [52]. Moreover, the temperature changes that affect the region of the nose, have been considered reliable to detect cognitive tasks. For example, nose temperature has been associated to mental workload, where in a driving test study simulator drives led to a higher subjective workload score and a greater nose temperature drop than real driving [53].

Mental workload increases with memory work, used in reading comprehension, and correlates to engagement tasks of sustained attention [4]. To create a biofeedback related to engagement in reading tasks, preliminary work showed mental workload can be used as an indication of engagement versus non-engagement, in a negative correlation between nose temperature and subjective immersion [39]. In reading tasks the temperature of the nose decreases when the reader is engaged, while increasing when not being engaged. In our preliminary study a correlation in engagement between the metrics of the heart and those of the temperature of the nose was found. When engaged, the skin temperature of the nose decreased (in red), and the temperature when non-engaged increased (in blue) (See Fig. 2). The metrics of heart showed that the HRV had higher and lower values in the non-engaging exercise than in the engaging exercise in accordance to the "Graph of psychophysiological interaction distinguished by the typology" from the Institute of HeartMath.

In synthesis, the metrics of nose temperature could be related to an engaged experienced of sustained attention related to the flow theory in reading tasks, as mental workload increases with memory work. It has been suggested that working memory is an indication of efficiency to sustain attention in multiple task-relevant representations, even when there is distracting irrelevant information [15]. For instance, Kintsch and Ericsson argue that working memory serves to hold a few concepts used

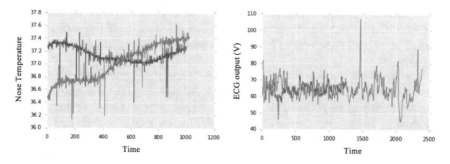

Fig. 2 Correlation of engagement (*red*) versus non-engagement (*blue*) between nose temperature (*left*) and HRV (*left*)

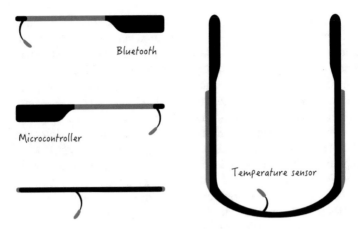

Fig. 3 Sketch of suggested smart-glasses to detect engagement in reading tasks

as cues to link what is stored in long-term memory [38]. Readers compose their own episodic structure for comprehension using the general knowledge to decode the written words and personal experience to give meaning [14]. Thus, for the biofeedback required to create an augmented narrative, the design of smart glasses that follow the reader's engagement with the metrics of memory workload reflected on the changes in nose temperature is proposed (See Fig. 3).

5 Sonic Assistance

The use of sonic assistance in literature is the result of studying the evolution of storytelling, from orality to literacy, with an emphasis on theatre [71]. When analyzing the medium's past, sound was found to remain as a constant component of storytelling. Thus, the proposed sonic augmentation looks into reintroducing components of orality to literature in a dual-coding across senses, where sound effects assist readers by addressing a channel other than the visual. Sound is used to fill in information gaps, with non-verbal information, allowing for an embodied experience. These sound effects act as positive disruptions awakening the mind to increase engagement.

Eric Havelock suggests the middle point between orality and literacy is in Attic theatre, which in itself arose from the need of Athenians in the sixth century to rediscover their own identity as a single city-state [24]. Attic theatre was a supplement for Homer, whose epics furnished the Greek identity, moral, politics and history. But Homer's tales were Ionic and Pan- Hellenic, and not part of the native dialect with which to address new Athenians. More importantly, Attic theatre gave birth to authors that were more producers than writers. They composed their vision, in the tension of oral and written communication by dictating to a literate assistant, and hearing it back from this assistant to verbally edit, retaining only the ear to compose. Havelock suggests the secret for the brilliance of Attic theatre is due to the tension caused by the transition from orality to literacy, which has not been repeated in history.

This tension is still visible in what Walter Ong calls residual oral cultures. For example, the acclaimed playwright William Shakespeare was part of these residual oral cultures. As such, it is unlikely that Shakespeare was involved in publishing his own plays, since his writing was meant to be spoken, and not read. In fact, John Marston, a contemporary playwright claimed: "*scenes invented to be spoken should not be enforced to the public for reading*" [57]. This is because Shakespeare's plays entail an interaction that is directly addressed to a particular audience, in a particular moment. However, generations of editors have added layers of silent emendations, holding meaning through grammatical punctuation, with commas and periods that set off clauses for the eye. These editorial revisions make Shakespeare more readable, since print is sophisticated and precise, but take away the rhetorical and auditory punctuation, disconnecting our embodied experience of theatrical narration [44].

With the evolution of storytelling, from orality to literacy, the perception of the narrative became purely visual, where the reader depicts the immediate external stimuli to the organism as neural representations, and relates it to internal portrayals encoded in memory. However, perception is not limited to only one sense. Events in everyday life are registered by more than one modality, integrating the different information from various sensory systems in a unified perception. For example, in Elizabethan plays not all sights of the action could be portrayed on stage, but even when these actions could not be seen, they all could be heard, making sound an important part of the narration. These off stage sounds were so important that packs of hounds and soldiers with their full arsenal were hired to produce authentic noises on cue, allowing for a multimodal perception.

Specific cues, such as sound effects or music fall into different dimensions, all of which need to be informational in nature to support off-the-moment events [40]. Even when the source is not in the field of vision, sound effects are powerful tools for stimulation. Audiences are able to recognize sounds by linking them to past sonic experiences, allowing them to mentally identify the source of each sound, and the significance it has to the narration. Even when sensory information is insufficient for the listener, the perceptual system still analyzes the situation taking into consideration previous knowledge acquired from the surrounding sonic-world [46].

In ecological psychology the physical nature of the sounding object, the means by which it has been set into vibration and the function it serves to the listener, is described to be perceived directly without any intermediate processing [50]. For example, studies show that listeners formulate the same cognitive organization based on the mechanics of the sound source: machine and electric device sounds, liquid sounds and aerodynamic sounds, even when these are sound-generated human-made illusions [28, 41]. Overall, sound has the ability to carry on non-verbal information. Sonic cues are able to structure perception by associating themselves to images and meanings, contributing for a significant experience, placing sound at the heart of interpretation [70].

5.1   Dual Coding Across Senses

Allan Paivio studied the unique human skill to deal simultaneously with verbal and non-verbal information [54]. In Pavio's dual-coding theory, the general assumption is that there are two classes of phenomena handled by separate cognitive subsystems: one for representation and processing of information, and the other for dealing with language. Language is a peculiar system that deals directly with speech and writing, while at the same time serves with a symbolic function with respect to non-verbal objects, events, and behaviours. For example, literature makes use of non-verbal information assistance adding illustrations to improve comprehension. In fact, students can perform better in reading comprehension test when there is a mix of text and images [32].

On the other hand radio drama, also called theatre of the mind, uses a dual-coding that addresses not the eyes, but the ears. Without any visual component, radio drama holds its meaning in the auditory dimension, depending on dialog, music, and sound effects to deliver enough information for the imagination. Here, sound effects round out the dialog, filling the absence of visual cues to convey meaning, and allow listeners to become part of the intricate moments of the drama. Sound is so powerful for the imagination that the auditory experience consents the absence of reason and logic. Listeners can lose structured thoughts to invite ideas that cannot be explained [72]. For example, in 1938 the radio play "War of the Worlds" caused hysteria amongst listeners who believed the Martians were invading. Imagery, also referred as listening with the mind, provided the radio audience in 1938 an aural visualization of extraterrestrial chaos, making them forget they were only tuning in to hear their regular program.

Regardless of the advantages of sonic non-verbal cues in radio drama, audiobooks have only retained the printed word. Even so, components of oral storytelling are still present. For instance, listening to an audiobook can make narrative texts that seem tedious to read, reveal the fullness of the literary work; especially if the work is narrated by a good actor [65]. Moreover, listening to what is being read can assist comprehension. One example is the modernist writer James Joyce, who experimented with aural reading to achieve comprehension. In Joyce's "Finnegans Wake" there is an acknowledgement of the importance of the sensory system, where understanding emerges from sound. The language developed by Joyce forces his readers to become aware that "Finnegans Wake" has to be pronounced, preferably out loud, to be able to understand the story [60]. To retrieve meaning readers needs to get into a rhythm that only can be achieved through orality.

In aural reading there is a redundancy of information, presented in a combination of auditory and visual channels, where readers process the same information twice. Language presented in these two modalities produces and enhanced memory recall [56]. Applied in multimedia learning, this verbal redundancy can facilitate the narration. When words are presented in both channels, learners are able to extract meaning from both modalities with no cognitive overload, since working memory and auditory working memory are independently processed [51].

In synthesis a dual-coding assists comprehension. In literature this is done through the visual channel with text and images, while in radio drama this is done through the auditory channel with speech and sound effects. Finally a dual-coding seems to be effective in a redundancy of verbal information across both channels: visual and auditory. However, there is an unexplored application using the potential of sound to carry non-verbal information mixed with the verbal information that the narrative text conveys.

In an augmented narrative verbal information, retrieved through the visual channel, is augmented with non-verbal information using sound effects in a dual-coding across senses. The non-verbal sonic information is used to technologically mediate the economy of attention, assisting the reading task. The framework depends on the mind-reading mechanisms, which literature simulates and capitalises on to mentally construct the storyworld [59]. It depends on the ability of the reader to find behavior patterns based on narrow slices of self experience [20]. Thus, an augmented narrative assists by giving more sensory information to the reader.

5.2 Sonic Disruption

Shakespeare and his plays hold the clue on how sound can assists readers: disruption. Shakespeare made a profound impact in the English language, refashioning oral expression through pen and paper. He has been credited for over 1700 original words, by changing nouns into verbs, verbs into adjectives, and by connecting words never before used together. However, Shakespeare's new words have a deeper impact into human conscience than just transforming oral expression. For example, in his play "Coriolanus", the main character returns to Rome for revenge. Menenius, well regarded by Coriolanus, is sent to persuade him to halt his crusade for vengeance. After sending Menenius back with no truce, Coriolanus recognizes: *"This last old man, whom with a crack'd heart I have sent to Rome, loved me above the measure of a father; nay,* **godded** *me, indeed"*. In this dialog, Shakespeare stimulates the reader by turning the noun God into a verb, where the new word 'godded' causes a grammatical disruption that triggers a P600 wave in the brain [68].

The P600 wave is a positive voltage variation peak in an electroencephalogram recording; a response found in the posterior part of the scalp that starts about 500 ms after hearing or reading an ungrammatical word. The wave reaches its maximum amplitude in around 600 ms, giving the reason for the name P600 [22]. More importantly, P600 is related to violations in the probability of occurrence of a stimulus, placing it in the P300 family [10]. This makes Shakespeare's new verbs unexpected stimulation rather than just grammatical disruptions.

The design of an augmented narrative involves translating Shakespeare's grammatical disruptions into sonic disruptions that act as assistive augmentations, favouring an embodied experience of literature with a dual-coding of information across senses, but also awakening the mind calling for the reader's attention. Here, sound is also used as spontaneous stimuli to technologically mediate attention when engagement is low, using the previously described smart glasses. Finally, it is important to mention disruption is not unique to Shakespeare's work. Alberto Moreiras, academic and cultural theorist, refers to literary disruptions as a "reversal of expectations", and argues that if properly managed, can prove more enabling and empowering to the story that was interrupted [34].

6 Augmented Narrative

Augmented Narrative is literature that assists readers to remember, feel, and imagine, triggering sound effects when engagement is low (See Fig. 4). The visual channel is enhanced by adding appropriate auditory information in a dual-coding across senses where information is perceived in an embodied experience. Here, interpretation of read actions is influenced by information available in the auditory channel. The augmented narrative operates in a cognitive ecology, following the body's reaction to the narrative text, technologically mediating attention and perception with multidimensional content.

Fig. 4 The concept of augmented narrative has 3 components: (*1*) Sound effects embedded in the narrative text, (2) eye tracker for input of reading location, (3) smart-glasses to measure engagement. Sonic assistance comes when two conditions are met: the eyes are over the span region, and engagement is low

6.1   Eye-Tracker

Reading skills vary from reader to reader, but eye tracking can serve as input of location, to know the area of the text the reader's gaze is, regardless of reading speed [6]. With the use of eye-tracking technology, the sonic augmentation is designed to work as an eye-driven-stimuli using an off-the-shelf eye-tracker technology to process the eye gaze, and link the sound to the perception of the narrative text. In this way individuals can read at their own pace, and when their gaze is in the 'span region' the sound is triggered. The eye-tracker tobii EyeX is used to design the eye-driven command during the reading task, in a transparent interaction where the reader is not distracted from the task to activate the sonic content (See Fig. 5). The "Gaze Track Plugin" developed by Augusto Esteves was used to connect the eye tracker to the Processing sketch. This plugin allows the tobii EyeX eye-tracker to trigger the embedded sound, marked in a span region. The eye gaze then instructs the sketch to pull the sound from a library of prerecorded .*wav* files. The eye tracker serves only as an input of location, leaving the command of assistance to the smart glasses. Thus, for the sound to be triggered two conditions need to be met: the gaze has to be over the span region and the nose temperature has to increase.

6.2   Engagement Levels

The smart-glasses create a biofeedback, giving the narrative text a cognitive component, allowing for a reciprocal communication where the reader perceives the narrative text, and the narrative text perceives the reader's level of engagement. The eye-driven sonic stimuli is only triggered if engagement is low, a sign that the reader is in need of assistance. Being triggered by involuntary reactions of the body, and not by the conscious mind, these sonic stimulations are set to disrupt prediction in

Fig. 5   Reading tablet with the tobii EyeX eye-tracker and engagement glasses

reading, calling for the readers attention. If the level of engagement is high, then there is no need for assistance, and the augmented narrative can leave the reader to continue the reading experience on her own, avoiding unnecessary disruption.

6.2.1 Measuring Engagement

In a first study we looked for the correlation between the changes in the temperature of the nose and reading engagement. We used our prototype of multimodal literature: (1) smart glasses with skin temperature to measure engagement; (2) Four original short stories from the series "Encounters" (*The Watchmaker, John, The Last Space, That Which Lives in the Attic*) in two modalities: sound and no-sound for control; (3) The tobii EyeX eye tracker mounted on the Dell Venue 11 Pro tablet. Eight volunteers (4 woman, 23–34 years) agreed to participate in testing the augmented narrative. The sample was based on the target readers for digital publishing demographics. All volunteers were given an explanation of the experimental setup and assigned two of the four short stories in a latin square design, one with sound and one with no-sound. Before they could start reading, each volunteer was calibrated for the nose temperature sensor and eye tracker. Volunteers took 2–6 min to read each short story depending on which short stories they were assigned. After each reading they were asked to fill in an immersion questionnaire [33] to measure subjective engagement for both short stories.

A Spearman's rank-order correlation was run to determine the relationship between reader's immersion levels and nose temperature. Combining the values of the two modalities, namely sound and no- sound, a negative correlation was found between immersion and temperature, which was statistically significant ($r_s(16) = -0.517$, $p = 0.04$) (See Fig. 6). Regarding the difference between modalities, we noticed subjects in the sound modality seemed to drop their nose temperature in more occasions than the no-sound modality 4:1, but did not find a statistical significance.

6.3   Sound

Through the analytical reflection that spontaneity is good, an augmented narrative carefully plans to be spontaneous, embedding sound that addresses particular passages that vanish with the passage itself. The narrative text is carefully edited with sound effects to stimulate the reader's mental simulation in an embodied perception of the story. The process begins by identifying sonic cues based on their capacity to effectively transmit non-verbal information relevant to the text, in an approach that fits the artistic process of writing fiction. These cues have been divided into five categories for assistance.

Fig. 6 Visualization of the negative correlation between nose temperature and engagement for the 8 participants in the 2 modalities (16 sets). Temperature decreases with higher engagement

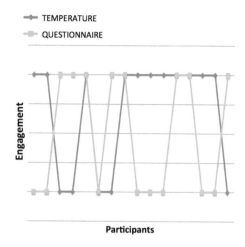

1. Redundancy. There are two sources of information, verbal and non-verbal that relate and complement each other creating an intermodal redundancy. For example when reading: *the phone rang*, the sound can complement the text by giving information about that phone. Is it a mobile, rotary dial or a touch-tone dialling phone?

2. Valid co-occurrences. To bring simultaneous sensory information of an event. For example, when the narrative talks of an explosion there is a synchronization of sound [5].

3. Inference. Writers give cues to their readers to infer situations and emotions, such as a blush or a sweaty hand to tell readers a certain character is in love. These cues can be given with sound, with a palpitating heart, where the reader can hear the intense heart beating of the character.

4. Differentiate. Sonic cues can be use to differentiate between parallel worlds or parallel times using a soundscape.

5. Ambience. The description of the story's setting is critical to understand the fictional world, as it helps to establish time and place. Sounds of an old and squeaky house could be the ambience of a horror story.

6.3.1   Measuring Sonic Assistance

To measure sonic assistance we looked for the best condition to increase reading comprehension and/or imagery between sonic non-verbal cues and sonic verbal redundancy. In this study an extract of "The Legend of Sleepy Hollow" was presented in three modalities: augmented with non-verbal sound effects, augmented with the audiobook taken from LibriVox for verbal redundancy, and without any sound for control. The short story was presented on the Dell Venue 11 Pro tablet computer mounted with the tobii EyeX eye tracker to fifteen volunteers (7 woman, 23–44 years) that agreed to participate in testing the augmented narrative.

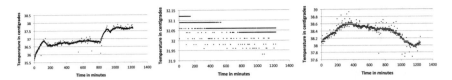

Fig. 7 Nose temperature samples for audiobook (*left*), control (*middle*), and sound effects (*right*)

All volunteers were given an explanation of the experimental setup and assigned one of the three modalities randomly. Before they could start reading, each volunteer was calibrated for the nose temperature sensor and eye tracker. Volunteers took 20 to 30 min to read the extract of the short story. After each reading we asked them to fill in an immersion questionnaire [33] to measure subjective engagement for both short stories and to retell [62] the story in order to be assessed in comprehension and imagery. The retell interview looked for the level of detail provided by each volunteer, comparing it to a previously prepared outline of five categories: characters, event details, climax, setting, and personal connections.

Results form a Kruskal-Wallis H test indicate that there was a statistically significant difference in nose temperature levels between the different modalities, $\chi^2(2) = 6.020$, $p = 0.049$, with a mean rank temperature level of 5.00 for the control group, 7.20 for non-verbal audio cues and 11.80 for participants exposed to audiobooks (See Fig. 7). We found no significant effects for immersion ($\chi^2(2) = 1.067$, $p = 0.587$) or comprehension ($\chi^2(2) = 2.226$, $p = 0.328$). However, were able to visually identify the sound effect modality was better for imagery. Sound appeared to change the temperature of the nose, positively and negatively when compared to the control modality, as this modality had only slight changes in temperature and average immersion. We observed sound effects seemed to enhance imagery, while verbal redundancy helped with comprehension. Even though comprehension was not significant, we noticed participants in the redundancy modality were more confident in the retell interview.

6.3.2 Conclusion

The concept of an augmented narrative follows a historical progression in a co-evolutionary process between literature and technology using familiar instruments and spaces to exploit the brain's natural strengths. Following Gutenberg's approach, where the success of print rests in preserving the quality of scripts, as well as solving production strains, we looked for an interaction that preserves the strengths of literature, but trying to open the reader's cognitive system to new possibilities when consuming it.

Preliminary studies suggest that the nose temperature could be a good indication for engagement levels. The term 'engagement' is proposed as it is part of the flow theory, allowing us to link immersion, more suitable for gaming, with attentional levels related to memory work found in the nose temperature. However, one limitation is attention overload. If the reader struggles retrieving the verbal information from the text, it could lead to stress, where the smart glasses would starts to work against the goal. In this case the disruption of sonic augmentation proved to only distract and frustrate the reader more. Moreover, if there is no textual information to relate the sound to, the cue becomes purposeless. Nevertheless, engagement could be found using the smart glasses, as in the study frustration seemed to lead to a much higher increase in nose temperature than engagement, by around four degrees.

The sonic assistive augmentation took most participants by surprise, since they did not know when the next sound would come. Regardless the disruption, we noticed participants were able to modify their biological minds and integrate the sonic information to the information of the narrative text. We suggest sound, as music and literature, is consumed across time. To understand the narrative, one needs to link all available information. We saw some indication of this in the debriefing session, even when participants could not immediately relate the sound to the text, as they progressed in the story they could remember and associate those sounds to what was read before. The sonic assistance seemed to encourage participants to imagine the setting of the story or even to correct their perception of the narrative. Some of them trusted more the recognizable sounds than the words, giving more weight to the assistive augmentation.

References

1. Angelotti M, Behnke RR, Carlile LW (1975) Heart rate: a measure of reading involvement. Res Teach Engl 192–199
2. Ashby J, Rayner K (2012) Reading in alphabetic writing systems: evidence from cognitive neuroscience. Neurosci Educ Good Bad Ugly 61
3. Benton M (2005) Reader-response criticism. Understanding children's literature: key essays from the second edition of the international companion encyclopedia of children's literature, 86
4. Berka C, Levendowski DJ, Lumicao MN, Yau A, Davis G, Zivkovic VT, Olmstead RE, Tremoulet PD, Craven PL (2007) Eeg correlates of task engagement and mental workload in vigilance, learning, and memory tasks. Aviat Space Environ Med 78(Supplement 1):B231–B244
5. Bertelson P, de Gelder B (2004) The psychology of multimodal perception. Crossmodal space and crossmodal attention, pp 141–177
6. Biedert R, Buscher G, Schwarz S, Möller M, Dengel A, Lottermann T (2010) The text 2.0 framework. In: Workshop on eye gaze in intelligent human machine interaction. Citeseer, pp 114–117
7. Cannon WB (1927) The james-lange theory of emotions: A critical examination and an alternative theory. Am J Psychol 106–124
8. Cohen B, Izard C, Facial Simons R (1986) physiological indexes of emotions in mother-infant interactions. In Psychophysiology, vol. 23, Soc Psychophysiol Res, (1010) Vermont Ave Nw Suite 1100. Washington, DC 20005:429–429
9. Corradini A, Antonietti A (2013) Mirror neurons and their function in cognitively understood empathy. Conscious Cogn 22(3):1152–1161

10. Coulson S, King JW, Kutas M (1998) Expect the unexpected: event-related brain response to morphosyntactic violations. Lang Cogn Process 13(1):21–58
11. Csikszentmihalyi M, Csikszentmihaly M (1991) Flow: the psychology of optimal experience, vol 41. Harper Perennial, New York
12. Eason SH, Goldberg LF, Young KM, Geist MC, Cutting LE (2012) Reader-text interactions: How differential text and question types influence cognitive skills needed for reading comprehension. J Educ Psychol 104(3):515
13. Ekman P, Levenson RW, Friesen WV (1983) Autonomic nervous system activity distinguishes among emotions. Sci 221(4616):1208–1210
14. Engle RW, Tuholski SW, Laughlin JE, Conway AR (1999) Working memory, short-term memory, and general fluid intelligence: a latent-variable approach. J Exp Psychol Gen 128(3):309
15. Ericsson KA, Kintsch W (1995) Long-term working memory. Psychol Rev 102(2):211
16. Fludernik M (2002) Towards a 'natural' narratology. Routledge
17. Folkerth W (2014) The sound of Shakespeare. Routledge
18. Gibson JJ. The senses considered as perceptual systems
19. Gibson JJ (2014) The ecological approach to visual perception: classic edition. Psychology Press
20. Gladwell M (2007) Blink: the power of thinking without thinking. Back Bay Books
21. Goldman AI (2005) 2 imitation, mind reading, and simulation. Perspectives on imitation: imitation, human development, and culture, 79
22. Gouvea AC. How to examine the p600 using language theory: what are the syntactic processes reflected in this component?
23. Graesser AC, McNamara DS (2011) Computational analyses of multilevel discourse comprehension. Top Cogn Sci 3(2):371–398
24. Havelock EA (1980) The oral composition of greek drama. Quaderni Urbinati di Cultura Classica, 61–113
25. Healy SD (1984) Boredom, self, and culture. Fairleigh Dickinson University Press
26. Herman D (2000) Narratology as a cognitive science. Image Narrative 1:1
27. Hollan J, Hutchins E, Kirsh D (2000) Distributed cognition: toward a new foundation for human-computer interaction research. ACM Trans Comput Hum Interact (TOCHI) 7(2): 174–196
28. Houix O, Lemaitre G, Misdariis N, Susini P, Urdapilleta I (2012) A lexical analysis of environmental sound categories. J Exp Psychol Appl 18(1):52
29. Hutchins E (2010) Cognitive ecology. Topics. Cogn Sci 2(4):705–715
30. Ioannou S, Ebisch S, Aureli T, Bafunno D, Ioannides HA, Cardone D, Manini B, Romani GL, Gallese V, Merla A. The autonomic signature of guilt in children: a thermal infrared imaging study
31. Ioannou S, Gallese V, Merla A (2014) Thermal infrared imaging in psychophysiology: potentialities and limits. Psychophysiology 51(10):951–963
32. Jalilehvand M (2012) The effects of text length and picture on reading comprehension of iranian efl students. Asian Soc Sci 8(3):329
33. Jennett C, Cox AL, Cairns P, Dhoparee S, Epps A, Tijs T, Walton A (2008) Measuring and defining the experience of immersion in games. Int J Hum Comput Stud 66(9):641–661
34. Kadir D (2006) Comparative literature in a world become tlon. Comp Crit Stud 3(1):125–138
35. Keller J, Bless H, Blomann F, Kleinböhl D (2011) Physiological aspects of flow experiences: skills-demand-compatibility effects on heart rate variability and salivary cortisol. J Exp Soc Psychol 47(4):849–852
36. Kiesling LL (2012) Mirror neuron research and adam smith's concept of sympathy: three points of correspondence. Rev Austrian Econ 25(4):299–313
37. Kintsch W (1988) The role of knowledge in discourse comprehension: a construction-integration model. Psychol Rev 95(2):163
38. Kintsch W, Patel VL, Ericsson KA (1999) The role of long-term working memory in text comprehension. Psychologia 42(4):186–198

39. Kunze K, Sanchez S, Dingler T, Augereau O, Kise K, Inami M, Tsutomu T (2015) The augmented narrative: toward estimating reader engagement. In: Proceedings of the 6th augmented human international conference. ACM, pp 163–164
40. Lebrecht J, Kaye D (1999) The art and technique of design, sound and music for the theatre
41. Lemaitre G, Heller LM (2013) Evidence for a basic level in a taxonomy of everyday action sounds. Exp Brain Res 226(2):253–264
42. Lewis D (1978) Truth in fiction. Am Philos Q 15(1):37–46
43. MacDonald RA, Hargreaves DJ, Miell D, Davidson JW, North AC (2002) Musical identities, vol 13. Oxford University Press, Oxford
44. MacLuhan M (2005) The effect of the printed book on language in the 16th century. Gingko Press
45. Mar RA, Oatley K (2008) The function of fiction is the abstraction and simulation of social experience. Perspect Psychol Sci 3(3):173–192
46. McAdams SE, Bigand EE (1993) Thinking in sound: the cognitive psychology of human audition. In: Based on the fourth workshop in the tutorial workshop series organized by the hearing group of the French acoustical society. Clarendon Press/Oxford University Press
47. McCraty R, Atkinson M, Tomasino D, Bradley RT (2009) The coherent heart: heart-brain interactions, psychophysiological coherence, and the emergence of system-wide order. Integr Rev 5(2):10–115
48. Mcluhan HM (2010) Understanding me: lectures and interviews. McClelland & Stewart
49. Merleau-Ponty M, Smith C (1996) Phenomenology of perception. Motial Banarsidass Publishe
50. Michaels CF, Carello C (1981) Direct perception. Prentice-Hall Englewood Cliffs, NJ
51. Moreno R, Mayer RE (2002) Verbal redundancy in multimedia learning: when reading helps listening. J Educ Psychol 94(1):156
52. Nakanishi R, Imai-Matsumura K (2008) Facial skin temperature decreases in infants with joyful expression. Infant Behav Dev 31(1):137–144
53. Or CK, Duffy VG (2007) Development of a facial skin temperature-based methodology for non-intrusive mental workload measurement. Occup Ergon 7(2):83
54. Paivio A (1978) A dual coding approach to perception and cognition. Modes of perceiving and processing information, pp 39–51
55. Pauletto S (2014) Film and theatre-based approaches for sonic interaction design. Digit Creativity 25(1):15–26
56. Penney CG (1989) Modality effects and the structure of short-term verbal memory. Mem Cogn 17(4):398–422
57. Pollard AW (2010) Shakespeare's fight with the pirates and the problems of the transmission of his text. Cambridge University Press
58. Purves AC, Rippere V. Elements of writing about a literary work—a study of response to literature
59. Rabinowitz PJ (1997) Before reading: narrative conventions and the politics of interpretation. The theory and interpretation of narrative series, ERIC
60. Rose PA (2012) The classical trivium: the place of thomas nashe in the learning of his time (by marshall mcluhan). Can J Commun 37:4
61. Ryan M-L (1991) Possible worlds, artificial intelligence, and narrative theory. Indiana University Press
62. Sadoski M (1983) An exploratory study of the relationships between reported imagery and the comprehension and recall of a story. Read Res Q 110–123
63. Schütz A (1951) Making music together: a study in social relationship. Soc Res 76–97
64. Scolari CA (2012) Media ecology: exploring the metaphor to expand the theory. Commun Theory 22(2):204–225
65. Shokoff J (2001) What is an audiobook? J Popular Cult 34(4):171
66. Sowa JF. Conceptual structures: information processing in mind and machine
67. Spacks PM (1995) Boredom: the literary history of a state of mind. University of Chicago Press

68. Thierry G, Martin CD, Gonzalez-Diaz V, Rezaie R, Roberts N, Davis PM (2008) Event-related potential characterisation of the shakespearean functional shift in narrative sentence structure. Neuroimage 40(2):923–931
69. Tribble E, Sutton J et al (2011) Cognitive ecology as a framework for shakespearean studies. Shakespeare Stud 39:94–103
70. Verstraete P. The frequency of imagination: auditory distress and aurality in contemporary music theatre
71. Walter O (1982) Orality and literacy. TJPress, London, The technologizing of the word
72. Weaver P (2012) Radio drama: a "visual sound" analysis of John, George and drew baby. PhD thesis, University of Central Florida Orlando, Florida
73. Zajonc RB (1985) Emotion and facial efference: a theory reclaimed. Science 228(4695):15–21
74. Zhao H (2012) Emotion-driven interactive storytelling. PhD thesis, Bournemouth University

Part II
Design for Assistive Augmentation

Design for Assistive Augmentation—Mind, Might and Magic

Ellen Yi-Luen Do

1 Introduction

In writing this preface to the topic of **Design for Assistive Augmentation**, I am taking the opportunity to draw on my own work and experience to shed light onto different aspects of assistive augmentation, while addressing particular aspects of the principles for design.

I argue that the design for assistive augmentation should take 3Ms into considerations—**Mind, Might, and Magic**.

1.1 *Definitions*

Before discussing the idea of designing for assistive augmentation, let's review the definition of the three keywords—design, assistive and augmentation.

Design is the creation or implementation of a plan, a system or an object, the process of manipulating both the form and function, to satisfy known constraints and achieve certain objectives. Design involves rationally the specification, analysis and solving of the problems, with reflective, even emotional sense-making, and improvisational creative actions. In the current context, we are considering Interaction Design (IxD), User Interface (UI) Design, User Experience (UX) Design, User-Centered Design (UCD), and Universal Design (UD).

E.Y.-L. Do (✉)
Keio-NUS CUTE Center, National University of Singapore, Singapore, Singapore
e-mail: ellendo@acm.org

E.Y.-L. Do
ATLAS Institute, University of Colorado Boulder, Boulder, USA

Assistive is an adjective often used together with technology, as in Assistive Technology (AT) to include devices and services, such as objects, equipment, software systems, applications and environments, designed or intended to aid or assist people with different abilities to perform in an activity or task, or to function, for work, health, and daily living in the world. This umbrella term encompasses ideas of accessible, assistive, adaptive, rehabilitative devices to promote greater independence, to enable or enhance functional capabilities, reduce difficulties, for participation and achievements.

Augmentation is the action or process to strengthen, to increase, to extend, to give rise, to accelerate, to boost, or to reinforce the value of the condition, in rate, amount or state. Specifically, we are concerned about the context of Augmented Human, in which the design of assistive technology, user interfaces and interactions aim to seamlessly integrate with sensory input and perception, as well as motions and behaviors in embodied interactions.

In the following section, let's discuss and link the idea of Assistive Augmentation (AA) to the concept of 3Ms—Might, Mind and Magic.

1.2 Mind, Might and Magic

As human beings, we observe, reflect, and act. We build tools to make and fix things, to gather food and resources, to maintain health, comfort and safety, to communicate, contribute, and belong to our communities. These assistive tools are augmentation to existing human capabilities. According to Maslow's theory of the hierarchy of human motivation [1], there are five levels of human needs—physiological, safety, love/belonging, esteem, and self-actualization, represented as a

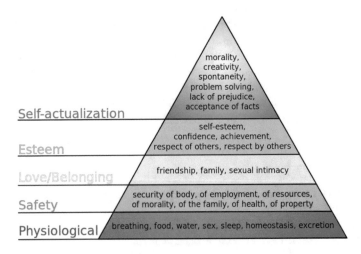

Fig. 1 Maslow's hierarchy of needs [2]

pyramid with more basic needs at the bottom as shown in Fig. 1. Whether we agree with the theory or not, we can acknowledge that there are universal human needs to be fulfilled, and we all possess different abilities to achieve different degrees of fulfilments. The question then, is—How can we design assistive augmentation to help advance and unlock human potentials to achieve different needs?

Earlier I have outlined the definitions of design, assistive and augmentation. Before describing some projects as illustrative examples in the next sections, I would like to argue that any design of successful assistive augmentation should take 3Ms into consideration: mind, might, and magic.

The **Mind** is about thinking, understanding, and planning. It's about ideas and making representations and decisions. The mind engages domain knowledge, retrieves relevant information, reasoning about actions, and stimulates creative thoughts.

The **Might** is the effects we have on the world, the efforts we engaged upon. The might represents the actual things we have accomplished using our power and strength. The might is the execution of activities, on objects, and environments.

Finally, the third M, **Magic**, concerns the development of design of the technology, to be well engineered to the extent that the form and function are seamlessly integrated into the fabric of daily lives, with wonderful, exciting effects. Magic is about making things work that appear to be fascinating, or enchanting. Science Fiction writer Arthur Clarke is famously quoted saying—*Any sufficiently advanced technology is indistinguishable from magic* [3].

To sum up, I am advocating the 3Ms in designing assistive augmentation— **Mind**, to observe before acting, to be thoughtful, and to be open-minded; **Might**, to consider the capacity, ability, efficacy, competency of people, and the technology; and finally, **Magic**, to have technology wonderfully blended in everyday life activities.

2 A Way of Working—Building Computational Tools

The premise of the Assistive Augmentation is that one can take on the challenge of integrating hardware and software technologies to improve user interfaces, and process contextual information to facilitate practical real-life applications.

Building computational tools is a way of working to solve problems and to innovate. Here, I am sharing my personal research journey to show that one can find inspirations from many different contexts, coming from very different disciplines towards the Design for Assistive Augmentation.

My interest in design computing started when I was as a graduate student at Harvard Graduate School of Design. Frustrated by the impoverished interface of advanced CAD software, I started programming and implemented sketching software to support design and eventually my Ph.D research, the *Right Tool at the Right Time—investigation of freehand drawing as an interface to knowledge based design tools* [4]. The important concept about building the Right Tool at the Right

Time is to detect what are the tasks at hand (reasoning about the context, antici-pating the need) and then to provide the appropriate support (triggering different functions, and knowledge-based systems).

Specifically, my research interests had spanned in several areas over the years: (1) computer aided design, especially sketch computing, (2) creativity and design cognition, including creativity support tools and design studies, (3) tangible and embedded interaction, and (4) computing for health.

2.1 Computer-Aided Design

Trained as an architect, my interest in design thinking led me to develop dia-grammatic and sketching interfaces for computer-aided architectural design. I asked, "What's in a hand-drawn design diagram that a computer should under-stand? [5]." To answer the question I studied design drawing and developed several computer-based sketching tools to explore and support design activities. For example, in Thinking with Diagrams [6] I examined how architects used dia-gramming and sketching as a tool to explore, discover, and develop ideas. The Design Sketches and Sketch Design Tools article [7] presents examples of various computational tools that can help designers in performing design tasks such as image retrieval, visual analysis, dimensional reasoning, and spatial transformation. Figure 2a shows that the VR Sketchpad system [8] transforms a 2D furniture layout sketch of a room into a 3D virtual reality for simulated walk-throughs to understand the use implications of the space. Figure 2b shows that the Design Evaluator system [9] analyzes and annotates spatial layout on the hospital floorplan based on design constraints for process flow or adjacency requirements that were defined in the program specification or rules and regulations in building codes.

I have also worked with students to develop tools for sketching in 3D [10], including annotation in virtual environments [11] to support collaboration and communication [12], sketch-to-building simulation [13], design evaluation [9],

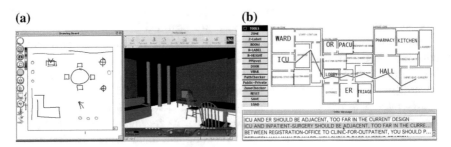

Fig. 2 **a** 2D sketch is transformed to 3D in VR Sketchpad system. **b** Design Evaluator system analyzes and annotates spatial layout in hospital floorplan based on design constraints for process flow and adjacency requirements

space making [14], and sketch-to-fabrication [15, 16]. I explored how sketching can serve as an interface to create or interact in 3D Virtual Reality environments [17].

2.2   Creativity and Design Cognition

To support creative design, I implemented diagram indexing and retrieval for shape-based finding of visual references [18], graphics interpreter of design actions [19], organization of a case-based design aid for architecture [20], as well as context recognition for designer's Ambiguous Intentions in their sketches [21]. In design studies and creativity and cognition, I have worked with students on the analysis of design [22], the role of physical objects and environment in creativity [23], and patterns of design process and collaboration [24].

2.3   Tangible and Embedded Interaction

My transition from 3D sketching to tangible and physical sketching follows the idea of "thinking with your hands" [25]. Tangible computing research engages designers to manipulate and experiment with embedded computing, to "sketch" with physical objects that comprise our built environment. Starting with the Navigational Blocks project [26] to navigate information space, my journey in tangible computing projects explore architectural form making [27], strategy games [28], energy awareness [29], construction kits [30], interactive furniture [31] and responsive environments [32].

For example, Fig. 3a shows the interaction with a tourist kiosk by picking up and rotating Navigational Blocks (Who, What, When, Where) would retrieve historical stories for different tourist attractions. Figure 3b shows that assembling and configuring Posey parts could constitute an animal puppet with replaceable body

(a)   **(b)**

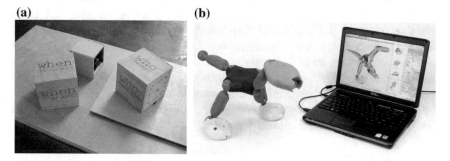

Fig. 3 a Interaction with Navigational Blocks to retrieve historical stories at tourist kiosk. **b** Dinosaur Pretend Play with configured Posey construction kit

parts that can be animated. These are examples of how tangible, physical objects with digital and contextual awareness can enhance the interaction for information retrieval and display.

2.4 Computing for Health Applications

I have worked on a variety of computing for health applications—using technology to encourage hand-washing in a smart patient room [33], a glove for spinal cord injury rehabilitation [34], an object identification tool for the visually impaired [35], a mobile health and robotic companion for children [36], and to promote employee active lifestyle [37], and the ClockReader [38], a system for automatically scoring the Clock Drawing Test that doctors use to screen for mild cognitive impairment. Instead of asking "What's in a design sketch that a computer should understand?" this project asks "What's in a dementia patient's drawing that a computer should understand?"

2.4.1 Electronic Clock Drawing Test (for Early Alzheimer's Disease's Detection)

Early detection is crucial for better planning and treatment managements of cognitive impairments such as Alzheimer's disease and other related disorders. Clock Drawing Test (CDT) is a well-established and commonly used paper-and-pencil screening instrument [39]. After each clock is drawn by a patient, the clinical staff then spend time to analyze the results by measuring and scoring each criterion (i.e., have all 12 numbers, long hand, short hand, and a center, numbers in the right locations, and sequence, etc.). We developed the ClockReader System with automated recording and analysis, time stamps and playback, so doctors and clinical staff could easily retrieve and monitor the progress of patients' cognitive impairment, exam the drawing process, and present the data in joint diagnosis sessions [40]. Figure 4a shows the ClockReader system interface with areas for drawing output, analysis, and monitoring. Figure. 4b shows the stylus pen and tablet computer setup for the ClockReader system. By implementing the pen and tablet interaction with simple user interface, we provide a computer environment that enables non-clinician or self-administered clock-drawing tests to be performed as frequently as needed, for easy comparison of past drawings collected overtime, as well as revealing time-based information such as the sequence of drawing [e.g., 1-3-6-9, or 10-11-12] and the idle time duration between each drawing marks (longer pause may indicate difficulty in recall).

(a)

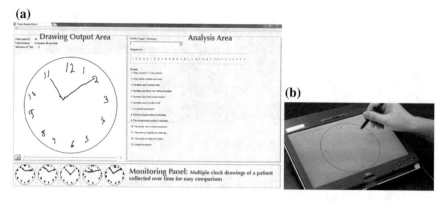

(b)

Fig. 4 **a** ClockReader system showing a patient's drawing (*left*), the automatic analysis and scoring (*right*), and past drawings displayed in the monitoring panel (*bottom*). **b** Stylus pen and tablet setup for ClockReader

(a)   **(b)**

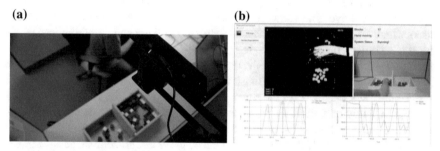

Fig. 5 **a** Digital Box and Block Test with Kinect setup above the desk. **b** DBBT screen showing detection of the positions of the fingers and the blocks, together with movement status

2.4.2 Digital Box and Block Test (Rehabilitation Independence)

Stroke is the leading cause of serious, long-term disability in the United States and worldwide [41]. Increasingly, we have seen stroke rehabilitation therapies conducted in patient's home. However, to perform the clinical assessments care providers still require patients to visit the clinic. The Digital Box and Block Test (DBBT) is a computational tool aims to help medical professionals record and assess rehabilitation progress of stroke patients with easy setup [42]. Figure 5a shows the Box and Block Test is augmented with a Kinect camera mounted above to record and perform analysis. As shown in Fig. 5b, the time-based movement data of the hand (including fingers and arm) can then be displayed and compared between sessions. Embedding this technology in the residential spaces could also help patients to relearn and recall how to use their arms, hands and fingers. With the system, care providers would be able to more precisely detect, track, and monitor patient's post-stroke functional motor improvements remotely.

Fig. 6 Vibrating motors of Mobile Music Touch cue which fingers to play the piano

2.4.3 Mobile Music Touch (Haptic Learning, New Skills and Rehabilitation)

An instrumented light-weight glove, the Mobile Music Touch [34, 43] is designed to facilitate passive haptic learning of piano playing by tapping corresponding finger for each key when the music is playing, so one can learn to play music while doing other tasks or on the move. The vibration motors outfitted on the Mobile Music Touch cue users which finger to use to play the next note. Figure 6 left and middle show a user's hand in a converted golf glove playing on a lighted keyboard, and the picture on the right shows a fingerless version of the music glove with a strap-on hardware box. The pilot study with students with no musical backgrounds shows that participants have fewer finger key mistakes for the songs that were cued with the glove than the ones without. A study of a short-term use of the glove with quadriplegic patients shows improved sensation and mobility for people with spinal cord injury [44]. This is a good example how a wearable device such as a music glove could facilitate passive music learning and engaging hand rehabilitation practice.

2.4.4 Tactile Teacher (Sensing Behaviors—Piano Learning, Experience Transfer)

A student often imitates the teacher's playing in terms of speed, dynamics, and fingering in a piano lesson. This learning model employs one's visual and audial perception for emulation, but it lacks the tactile sensation, an important component of piano playing. To investigate how we can convey the tactile sensations of the teacher's keystrokes to signal the student's corresponding fingers, we implemented Tactile Teacher, an instrumented fingerless glove to detect finger taps on hard surfaces [45]. Recognizing that finger taps generate acoustic signals and cause vibrations, after testing on several different sensor placements and orientations (Fig. 7 left), we embedded three vibration sensors on the glove, and use machine learning algorithms to analyze the data from the sensors. After a brief training procedure, this prototype (Fig. 7 right) can accurately identify single finger tap in a very good performance at above 89% accuracy, and two finger taps resulted in accuracy around 85%. Wouldn't it be nice if Tactile Teacher can capture the piano playing techniques

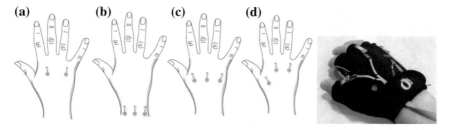

Fig. 7 Four configurations tested to determine the optimal sensor placements and orientations, resulting in an instrumented (configuration d) glove with vibration sensors in Tactile Teacher

from virtuoso piano players and then transfer the tactile sensations to learners through Mobile Music Touch? This also shows the potential of capturing other finger-based fine motor skills for training and rehabilitation in the future.

2.5 Things That Think, Spaces That Sense and Places That Play

Over the years, the focus of my work has been applying design computing and human-centered computing knowledge to investigate and implement the vision of *Things that Think, Spaces that Sense and Places that Play*—a smart living environment in which computing technologies embedded in the built environment (e.g., objects, furniture, building, and space) support everyday happy healthy living.

My recent work further explored the ideas of creating experience media and interactive computing projects towards smart living environments. Specifically, at the Keio-NUS CUTE Center we explore the idea of "Creating Unique Technology for Everyone" through the use of Connective, Ubiquitous Technology for Embodiments, in key areas of tangible interaction, augmented learning, and embodied experience [46].

3 Creating Unique Technology for Everyone

Design and Human-Computer Interaction are crucial components of information technologies in daily life and they color our experience of computation and communication. As transdisciplinary researchers and designers, we have a mission to pursue the vision of Creating Unique Technology for Everyone through the use of Connective, Ubiquitous Technology for Embodiments. Here I will describe a couple projects from CUTE Center that demonstrate the idea about how mobile, ubiquitous, or physical and tangible computing can be used to augment or enhance human abilities and experiences.

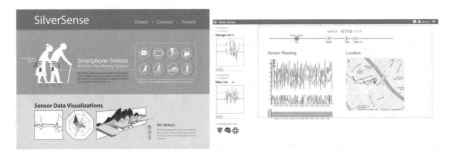

Fig. 8 SilverSense mobile application provides various data visualization of behaviors of seniors (*left*) such as movement and location info through time (*right*)

3.1 Sensorendipity and SilverSense (Mobile Phone, Behavior Monitoring)

SilverSense is a smartphone-based activity monitoring system for elderly senior citizens [47]. Utilizing a smartphone's built-in sensors (e.g., movement, location, light and sound levels), SilverSense uses the sensor data to facilitate the detection of old-age problems, such as dementia and falls, stores the sensor data for caregivers and family members to access and visualize the activity history to facilitate monitoring and better life style management. A collaboration with People's Association Active Ageing Council, the project aims to provide convenient, non-intrusive monitoring of elderly seniors' wellbeing, while connecting their family members and caregivers through a user-friendly interface. SilverSense is powered by Sensorendipity [48], a smartphone-based web-enabled sensor platform developed to facilitate smartphone sensors to be used easily by web developers to develop real-time web applications. Figure 8 shows that the SilverSense mobile app provides sensor data visualization as activity monitoring systems, and a display interface for the movement and location data.

3.2 SilverTune NinjaX (Transformable Toy for Music Therapy)

Music therapy is increasingly conducted in the health care and clinical settings such as rehabilitation centers and nursing homes to assist older adults with physical disabilities or mental impairments due to dementia or stoke. SilverTune NinjaX (Fig. 9a) is a smart assistive device based on Ninja Track [49], a collaboration between CUTE Center and Nanyang Polytechnic, with interviews of music therapists and occupational therapists from KTPH Geriatric Centre, AWWA Rehab & Elderly Care Centre, and AWWA Dementia Day Care Centre. While Ninja Track for Game was modified, and incorporated into a first-person "fishing/fighting" game Reel Blade [50, 51] as shown in Fig. 9b, Ninja Track for Music was adopted and

(a)   **(b)**

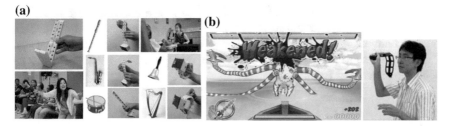

Fig. 9 **a** SilverTune NinjaX device in foldable and flexible form with buttons, that can be configured to play sounds of different musical instruments (flute, saxophone, hand bell, drum stick and harp roll). **b** The sword or fish reel game controller game Reel Blade, also an extension of Ninja Track

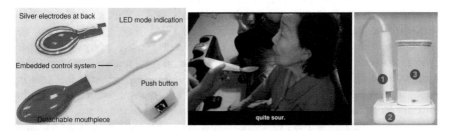

Fig. 10 (*Left*) Taste+ spoon prototype, (*middle*) study participant confirm the sour taste, and (*right*) Virtual Lemonade system with (1) sensor, (2) communication, and (3) simulation

revamped in customizable configurations as to several types of musical instruments with audios and play interactions to cater for different preferences and therapeutic requirements, in both individual and group settings, to quantitatively record therapeutic data, and analyze performance to give multi-modal feedback to both the elderlies and the therapists.

3.3 Taste+ (Digital Stimulation for Taste Enhancement)

Taste+, a winning entry of the inaugural Design Challenge hosted by Stanford Longevity Center [52, 53], is a spoon with built-in electronic control to enhance sourness and saltiness digitally for an elderly person's taste of food, without adding chemical flavoring ingredients to compensate for their diminishing sense of taste due to old age or cognitive impairments. When the tongue touches the two silver electrodes at the bottom of the spoon, the taste sensations of food and beverages can be enhanced (to be saltier or more sour), potentially reducing a person's salt intake. Borrowing the metaphor of the multi-color ballpoint pen color switching operation, one can push a button to switch taste with corresponding color (ocean blue salty or lime green sour) [54]. Our recent Virtual Lemonade system [55] further explores the opportunity of sensing and teleporting the color and corresponding PH value of a glass of lemonade to a customized tumbler to virtually simulate these properties with plain water. Figure 10 shows the Taste+ spoon design with electrodes, and push button to switch between the different tastes with corresponding LED light colors (left), a study participant stating the spoon is quite sour (middle), and the Virtual Lemonade system with three main components: (1) the lemonade sensor, (2) the communication protocol, and (3) a customized tumbler, acting as the lemonade simulator (right).

3.4 AmbioTherm (Thermal and Ambient Environment for Presence Enhancement)

AmbioTherm is a wearable accessory for Head Mounted Displays (HMD) that provides thermal and wind stimuli to simulate real-world environmental conditions, to enhance the sense of presence in Virtual Reality (VR). With an Ambient Temperature Module attached to the user's neck, and a Wind Simulation Module placed in front of the user's face with fans, the Control Module utilizing Bluetooth communication can provide wind and thermal stimuli for VR environments such as a snowy mountain and a hot desert. Participants of the study reported that wearing AmbioTherm significantly improves the sensory and realism factors, contributing towards an enhanced sense of presence when compared to traditional VR experiences [56]. Figure 11 shows the AmbioTherm setup that includes a Head Mounted Display, two servo motors-connected fans, and the Peltier elements attached on the back of the neck, all connected to a Microcontroller and Bluetooth Interface.

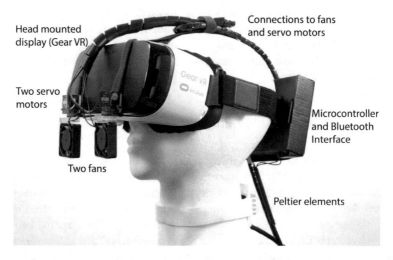

Fig. 11 AmbioTherm gives people the sensation of being in a hot desert, or a snowy mountain by providing thermo module that increases the temperature for heat, and two fans controlled by servo motors that would change wind directions to create the active motions in a cooler temperature ambient environment

4 Discussion and Future Work

Earlier in this article I propose the idea of 3Ms—**M**ind, **M**ight and **M**agic as design principles for Assistive Augmentation. I argue that building computational tools is a way of working, to build objects to think with, to implement innovation, and to facilitate creative engagements, using tangible and embedded interaction, for health applications. I advocate the environment for creativity to be a lab for making things and the aspiration to **C**reating **U**nique **T**echnology for **E**veryone through the use of **C**onnective, **U**biquitous **T**echnology for **E**mbodiments.

Then, you might ask, what kind of Design for Assistive Augmentation shall we work on? Let me provide you with some food for thought here.

As illustrated in Fig. 12, Nakakoji observed that people employ three types of physical tools: (1) dumbbell, (2) running shoes, and (3) skis, to help improve or enhance the performance of their physical activities. She suggested that researchers should take this analogy into consideration when evaluating different creativity support systems [57]. For example, dumbbells help people develop muscles. Once muscles are developed, they can be used for other physical activities, not just for lifting dumbbells. Running shoes, on the other hand, can help runners run faster or more comfortably. People can run without wearing the running shoes, but wearing

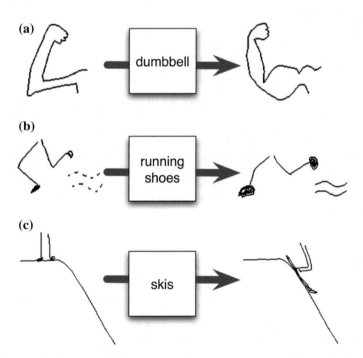

Fig. 12 Three types of tool support—**a** dumbbell to train abilities, **b** running shoes to enhance performance when wearing, **c** skis that enable new experience

the running shoes may result in better running experience. Finally, skis enable people to ski. Skiing is a new kind of experience that cannot happen without wearing the skis.

As we engage in the Design for Assistive Augmentation, let's remember to ask ourselves the following question: *Are we developing dumbbells to help people to build muscles, running shoes to help people run faster, or skis to enable people to ski?*

Since tools were designed for different purposes, therefore, their roles and effects should be considered accordingly in the research, development and evaluation process. While we can certainly find examples of tools that simultaneously embody multiple aspects, it's important for us to be aware of the differences among them, and not to arbitrary adopt a research design or evaluation framework that might be appropriate for one to the others. I also would like to encourage everyone to work on creating new type of physical and computational dumbbells, running shoes and skis to augment human capabilities.

This section focuses on the topic area of "**Design for Assistive Augmentation**". The succeeding three chapters in this section include (a) Designing Augmented, Domestic Environments to Support Ageing in Place, (b) Sensory Conversation: An Interactive Environment to Augment Social Communication in Autistic Children, and (c) FingerReader: A Finger-Wearable Assistive Augmenter. These three projects are good examples of the idea of "Design for Assistive Augmentation", because they demonstrated the support for real-world applications, from understanding the context and the physical environment to support the challenges of capturing relevant sensory information.

Acknowledgements This material is based upon research work supported by the US National Science Foundation under Grant SHB 1117665, DUE 0127579, IIS 96-19856, Emory ADRC/ACTSI Pilot Grant, Korea Institute for Advancement of Technology (KIAT)'s Global Industry-Academia Cooperation Program, Georgia Institute of Technology's Health System Institute, Strategic Energy Initiative, and Center for Music Technology, Atlanta's VA Center, Emory Center for Neurodegenerative Disease and Alzheimer's Disease Center, Pennsylvania Infrastructure Technology Alliance (PITA), People's Association (Singapore), and the National Research Foundation, Prime Minister's Office, Singapore under its International Research Centre in Singapore Funding Initiative. Any opinions, findings, and conclusions or recommendations expressed in this material are those of the author and do not necessarily reflect the views of the funding agencies.

References

1. Maslow AH (1943) A theory of human motivation. Psychol Rev 50(4):370–396. http://doi.org/10.1037/h0054346
2. Finkelstein J. Maslow's hierarchy of needs. https://commons.wikimedia.org/wiki/File:Maslow's_hierarchy_of_needs.png, https://creativecommons.org/licenses/by-sa/3.0/legalcode
3. Wikipedia. Arthur Clarke's three laws. https://en.wikipedia.org/wiki/Clarke%27s_three_laws

4. Do EY-L (1998) The right tool at the right time—investigation of freehand drawing as an interface to knowledge based design tools. Ph.D dissertation, Georgia Institute of Technology, GVU Center Technical Reports [536]. https://smartech.gatech.edu/handle/1853/3482

5. Do EY-L (1995) What's in a diagram that a computer should understand? In: Tan M, Teh R (eds) The global design studio, proceedings of the sixth international conference on computer aided architectural design futures (CAAD Futures 95), National University of Singapore, pp 469–482. ISBN:9971-62-423-0. http://papers.cumincad.org/cgi-bin/works/Show?a56e

6. Do EY-L, Gross MD (2001) Thinking with diagrams in architectural design. In: Artificial intelligence review, vol 15. Kluwer Academic Publishers, Dordrecht, The Netherlands, pp 135–149. https://link.springer.com/article/10.1023/A:1006661524497

7. Do EY-L (2005) Design sketches and sketch design tools. In: Nakakoji K, Gross MD, Candy L, Edmonds E (eds) KBS—knowledge based systems, vol 18. Elsevier Publisher, pp 383–405. http://doi.org/10.1016/j.knosys.2005.07.001. Accessed 11 Aug 2005.

8. Do EY-L (2001) VR sketchpad: creating instant 3D worlds by sketching on a transparent window. In: de Vries B, van Leeuwen JP, Achten HH (eds) CAAD futures 2001, July 2001. Kluwer Academic Publishers, Eindhoven, The Netherlands, pp 161–172. http://doi.org/10.1007/978-94-010-0868-6_13

9. Oh Y, Gross MD, Do EY-L (2004) Critiquing freehand sketches: a computational tool for design evaluation. In: Gero J, Knight T (eds) Visual and spatial reasoning in design III [VR'04], 22–23 July 2004. MIT, pp 105–120

10. Jung T, Gross MD, Do EY-L (2001) Space pen, annotation and sketching on 3D models on the internet. In: de Vries B, van Leeuwen JP, Achten HH (eds) CAAD futures 2001, July 2001. Kluwer Academic Publishers, Eindhoven, The Netherlands, pp 257–270

11. Jung T, Gross, MD, Do EY (1999) Immersive redlining and annotation of 3D design models on the web. In: Augenbroe G, Eastman C (eds) Computers in building (June, '99) proceedings of the CAAD futures '99 conference, pp 81–98

12. Jung T, Gross MD, Do EY-L (2002) Annotating and sketching on 3D web models. In: International conference on intelligent user interfaces (IUI), 13–16 Jan 2002. ACM Press, San Francisco, pp 95–102

13. Do EY-L, Gross MD (2004) Let there be light! Knowledge-based 3-D sketching design tools. In: Jabi W, Pinet C (eds) IJAC Int J Archit Comput (Multi-Science Publishing Co Ltd) 2 (2):211–227(17). http://doi.org/10.1260/1478077041518647

14. Lee M-C, Do EY-L (2003) Space maker: creating space by sketching it. In: Annual conference of ACADIA, association of computer aided design in architecture, 23–26 Oct 2003. Ball State University, pp 311–323

15. Oh Y, Johnson G, Gross MD, Do EY-L (2006) The designosaur and the furniture factory. In: International conference on design computing and cognition (DCC 06), 10–12 July 2006, Eindhoven, Netherlands

16. Johnson G, Gross MD, Do EY-L, Hong J (2012) Sketch it, make it: sketching precise drawings for laser cutting. In: CHI EA 2012, pp 1079–1082. http://doi.acm.org/10.1145/2212776.2212390

17. Do EY-L (2011) Sketch that scene for me and meet me in cyberspace. In: Wang X, Tsai JJ-H (eds) Collaborative design in virtual environments (chapter 11). Springer, pp 121–130. ISBN: 978-94-007-0604-0

18. Gross MD, Do EY-L (1995) Diagram query and image retrieval in design. In: 2nd IEEE international conference on image processing, vol 2. IEEE Computer Society Press, Washington, DC, pp 308–311

19. Do EY-L (2001) Graphics interpreter of design actions: the GIDA system of diagram sorting and analysis. In: de Vries B, van Leeuwen JP, Achten HH (eds) Computer aided architectural design futures (CAAD Futures 2001), July 2001. Kluwer Academic Publishers, Eindhoven, The Netherlands, pp 271–284

20. Zimring C, Bafna S, Do EY-L (1996) Structuring cases in a case-based design aid. In: Vanegas J, Chinowsky P (eds) Third congress on design computing, Anaheim, A/E/C'96. American Society of Civil Engineers (ASCE), pp 308–313

21. Gross MD, Do EY-L (1996) Ambiguous intentions—a paper-like interface for creative design. In: Proceedings of ninth annual symposium for user interface software and technology (UIST 96), pp 183–192

22. Cai H, Do EY-L, Zimring CM (2010) Extended linkography and distance graph in design evaluation: an empirical study of the dual effects of inspiration sources in creative design. J Design Stud 31(2):146–168. http://doi.org/10.1016/j.destud.2009.12.003

23. Williams J, Do EY-L (2009) Locomotion storytelling: a study of the relationship between kinesthetic intelligence and tangible objects in facilitating preschoolers creativity in storytelling. In: ACM creativity and cognition, Oct 27–30, Berkeley, pp 415–416. http://doi.org/10.1145/1640233.1640328

24. Lee S, Ezer N, Sanford J, Do EY-L (2009) Designing together while apart: the role of computer-mediated communication and collaborative virtual environments on design collaboration. In: IEEE systems, mans and cybernetics, San Antonio, Oct 11–14, pp 3693–3698

25. Do EY-L, Gross MD (2007) Environments for creativity—a lab for making things. In: Shneiderman B, Fischer G, Giaccardi E, Eisenberg M (eds) Creativity and cognition. ACM Press, New York, pp 27–36. http://doi.org/10.1145/1254960.1254965

26. Camarata K, Do EY-L, Johnson BR, Gross MD (2002) Navigating information space with tangible media. In: International conference on intelligent user interfaces (IUI), 13–16 Jan 2002. ACM Press, San Francisco, pp 31–38. http://dl.acm.org/citation.cfm?id=502725

27. Eng M, Camarata K, Do EY-L, Gross MD (2006) FlexM: designing a physical construction kit for 3D modeling. In: Cheng N, Pinet C (eds) IJAC Int J Archit Comput (Multi-Science Publishing Co Ltd) 4(2):27–47. http://doi.org/10.1260/1478-0771.4.2.27

28. Wu A, Joyner D, Do EY-L (2010) Move, beam, and check! Imagineering tangible optical chess on an interactive tabletop display. In: ACM Computers in Entertainment (CIE), vol 8 (3), Article 20 (15 pp.). http://dx.doi.org/10.1145/1902593.1902599

29. Camarata K, Do EY-L, Gross MD (2006) Energy cube and energy magnets. In: Cheng N, Pinet C (eds) IJAC—Int J Archit Comput 4(2):49–66. http://doi.org/10.1260/1478-0771.4.2.49

30. Weller MP, Do EY-L, Gross MD (2008) Posey: instrumenting a poseable hub and strut construction toy. In: Tangible and embedded interaction (TEI'08), Feb 18–20, Bonn, Germany, pp 39–46. http://portal.acm.org/citation.cfm?id=1347402

31. Oh Y, Camarata K, Weller MP, Gross MD, Do EY-L (2008) TeleTables and window seat: bilocative furniture interfaces. In: Theng Y-L, Duh HBL (eds) Ubiquitous computing: design, implementation, and usability. Hershey: Information Science Reference, 2008 (chapter 11), pp 160–171. ISBN-13: 978-1-59904-693-8 (hardcover); ISBN-13: 978-1-59904-695-2 (e-book)

32. Camarata K, Gross MD, Do EY-L (2003) A physical computing studio: exploring computational artifacts and environments. In: Cheng NY-W, Do EY-L (eds) Int J Archit Comput (Multi-Science Publisher, UK) 1(2):169–190. http://doi.org/10.1260/147807703771799166

33. Do EY-L (2009) Technological interventions for hand hygiene adherence—research and intervention for smart patient room. In: Tidalfi T, Dorta T (eds) CAAD futures, 17–19 June 2009, Montreal, Canada, pp 303–313

34. Huang K, Starner T, Do E, Weinberg G, Kohlsdorf D, Ahlrichs C, Leibrandt R (2010) Mobile music touch: mobile tactile stimulation for passive learning. In: CHI'10 proceedings of the 28th international conference on human factors in computing systems, pp 791–800. http://doi.org/10.1145/1753326.1753443

35. Lawson MA, Do EY-L, Marston JR, Ross DA (2011) Helping hands versus ERSP vision: evaluating the effectiveness of two wearable object recognition technologies. In: Stephanidis C (ed) Human-computer interaction thematic area, 9–14 July 2011, Orlando, in HCI International 2011: Posters, Part I, HCII 2011, CCIS 173, pp 383–388, 2011. Springer Berlin, Heidelberg 2011. https://www.hcii2011.org/

36. Lu S-C, Blackwell N, Do EY-L (2011) mediRobbi: an interactive companion for pediatric patients during hospital visit. In: Health and well-being applications, Tuesday, 12 July 2011:

08:00–10:00, Orlando, in HCI International 2011, vol 2, LNCS_6762, 2011. Springer, Berlin, Heidelberg, pp 547–556

37. Kim H, Swarts M, Lee S, Do EYL (2010) HealthQuest: technology that encourages physical activity in the workplace (2010). In: ICOST—international conference on smart homes and health telemetics, Seoul, June 22–24, pp 263–266. http://icost2010.org/

38. Kim H, Cho YS, Guha A, Do EY-L (2010) ClockReader: investigating senior computer interaction through pen-based computing. In: CHI workshop on senior-friendly technologies: interaction design for the elderly, 10 Apr 2010, Atlanta, GA, USA, pp 30–33

39. Agrell B, Ove D (1998) in Age and Ageing (1998), The clock-drawing test. 27:399–403 (Republished Age Ageing. 41(Suppl 3):iii41–5 (2012). http://doi.org/10.1093/ageing/afs149.)

40. Kim H, Hsiao C-P, Do EY-L (2012) Home-based computerized cognitive assessment tool for dementia screening. J Ambient Intell Smart Environ (JAISE IOS Press) 429–442. http://doi.org/10.3233/AIS-2012-0165

41. Stroke Statistics, Internet Stroke Center. http://www.strokecenter.org/patients/about-stroke/stroke-statistics/ US statistics from U.S. Centers for Disease Control and Prevention, retrieved July 2nd, 2017

42. Zhao C, Hsiao C-P, Davis N, Do EY-L (2013) Tangible games for stroke rehabilitation with digital box and blocks test. In: CHI EA'13 CHI'13 extended abstracts on human factors in computing systems. ACM, New York, NY, USA, pp 523–528. http://dl.acm.org/citation.cfm?id=2468448

43. Huang K, Do EY-L, Starner T (2008) PianoTouch: a wearable haptic piano instruction system for passive learning of piano skills. In: 12th IEEE international symposium on wearable computers ISWC 2008, pp 41–44, Sep 28–Oct 1, Pittsburgh. http://www.iswc.net/, http://doi.ieeecomputersociaety.org/10.1109/ISWC.2008.4911582

44. Markow T, Ramakrishnan N, Huang K, Starner T, Eicholtz M, Garrett S, Profita H, Scarlata A, Schooler C, Tarun A, Backus D (2010) Mobile music touch: vibration stimulus as a possible hand rehabilitation method. In: Proceedings of the 4th international pervasive health conference, March 2010, Munich, Germany

45. Li R, Wang Y, Hsiao C-P, Davis N, Hallam J, Do E (2016) Tactile teacher: enhancing traditional piano lessons with tactile instructions. In: Proceedings of CSCW (computer supported collaborative work), Feb 27–March 2, 2016. ACM Press, NY, pp 329–332. http://doi.org/10.1145/2818052.2869133

46. Do EY-L (2005) Creating unique technology for everyone. In: ASEAN CHI symposium'15 proceedings of the ASEAN CHI symposium'15, Seoul, Republic of Korea, 18–23 April 2015. ACM, New York, NY, pp 42–45. http://dl.acm.org/citation.cfm?id=2780363

47. Tan A (2014) New app to help seniors live more independently, 13 April 2014, pp 16, home section, Sunday Times, The Strait Times, Singapore. http://www.straitstimes.com/singapore/health/new-app-designed-by-nus-researchers-to-help-seniors-live-more-independently

48. Lu W, Sun CC, Bleeker T, You Y, Kitazawa S, Do EY-L (2014) Sensorendipity, a real time web-enabled smartphone sensor platform for idea generation and sensor platform. In: International symposium of Chinese CHI (Chinese CHI'14), Toronto, 26–27 Apr 2014, pp 11–18. http://dl.acm.org/citation.cfm?id=2592238. http://doi.org/10.1145/2592235.2592238

49. Katsumoto Y, Tokuhisa S, Inakage M (2013) Ninja track: design of electronic toy variable in shape and flexibility. In: International conference on tangible, embedded and embodied interaction, 10–13 Feb 2013. ACM, Barcelona, Spain, pp 17–24. http://doi.org/10.1145/2460625.2460628

50. Feit D (2015) This is the sword-slash-fishing reel you've been waiting for. In: Culture, Wired, 09.18.15, #INDIE GAMES #REEL BLADE #TOKYO GAME SHOW 2015. https://www.wired.com/2015/09/fishing-sword-transformer/ or https://web.archive.org/web/20160825164122/, http://www.wired.com/2015/09/fishing-sword-transformer/

51. (2015) Reel blade: battle of the high seas. http://gamelab9.wixsite.com/reelblade

52. Stanford (2014). http://longevity3.stanford.edu/design-challenge-winners-announced/

53. NUS News (2014) Electronic spoon helps the elderly age tastefully, Published: 16 May 2014, Category: Highlights. https://web.archive.org/web/20160517211857/, http://news.nus.edu.sg/highlights/7686-electronic-spoon-helps-the-elderly-age-tastefully

54. Ranasinghe N, Lee K-Y, Suthokumar G, Do EY-L (2014) The sensation of taste in the future of immersive media. In: ImmersiveMe'14 proceedings of the 2nd ACM international workshop on immersive media experiences 07–07 November 2014, Orlando, Florida, USA. ACM, New York, NY, USA, pp 7–12. http://doi.org/10.1145/2660579.2660586

55. Ranasinghe N, Jain P, Karwita S, Do EY-L (2017) Virtual lemonade: let's teleport your lemonade! In: Proceedings of the 11th international conference on tangible, embedded and embodied interactions, (TEI 2017), 20–23 March 2017, Yokohama, Japan, pp 183–190 (41/151 = 27%). http://dx.doi.org/10.1145/3024969.3024977

56. Ranasinghe N, Jain P, Karwita S, Tolley D, Do EY-L (2017) Ambiotherm: enhancing sense of presence in virtual reality by simulating real-world environmental conditions. In ACM human factors in computing conference—CHI'17 papers and notes, 06–11 May 2017, Denver, CO, USA, 2017. ACM, pp 1731–1742. ISBN: 978-1-4503-4655-9/17/05. http://dx.doi.org/10.1145/3025453.3025723

57. Nakakoji K (2006) Meanings of tools, support, and uses for creative design processes. In: The proceedings of international design research symposium'06, Nov 2006. CREDITS Research Center, Seoul, Korea, pp 156–165. ISBN: 89-950046-4-9

Designing Augmented, Domestic Environments to Support Ageing in Place

Jeannette Durick and Linda Leung

1 Introduction

[S]tereotypes are useful for camouflaging the social arrangements which we impose upon the aged members of our society. As the unspoken assumptions upon which 'scientific' theories of ageing are constructed, they become doubly dangerous, being mindfully or inadvertently employed to determine the fate of fellow human beings [24].

The last two decades have seen exponential growth in research areas related to designing technology for older adults. Much of this work has emphasised technology-centric approaches that focus on assistive technologies and products, specifically designed for the 'elderly', and appears to be driven by governments' concerns about ageing populations, as well as enduring, negative social attitudes regarding older adults. An unfortunate by-product of such work is that, often, it is built upon assumptions about older adults and designed for stereotypes, rather than real-life users or relevant personas, and thereby, potentially risks the wellbeing of intended users. However, some recent studies into technology and ageing have been based upon human-centred perspectives and incorporated, for example, user-centred design approaches in order to better understand the individual experiences of older adults and their requirements. Such work includes contributions from the fields of gerontology (e.g., [23, 33]), gerontechnology (e.g., [25, 44, 45]), Human-Computer Interaction (HCI, e.g., [22, 48, 51, 57]) and Computer-Supported Cooperative Work (CSCW, e.g., [12, 13, 49]).

J. Durick (✉) · L. Leung
University of Technology Sydney, Ultimo, NSW, Australia
e-mail: jeannette.durick@uts.edu.au

L. Leung
e-mail: linda.leung@uts.edu.au

© Springer Nature Singapore Pte Ltd. 2018
J. Huber et al. (eds.), *Assistive Augmentation*, Cognitive Science
and Technology, https://doi.org/10.1007/978-981-10-6404-3_7

Australia is just one of many developed countries whose Government has suggested that its ageing population will inevitably lead to (i) stress on healthcare and welfare systems, i.e., due to increased chronic illness, and (ii) a dwindling economy (e.g., [40]). However, various gerontologists and sociologists, such as Harvey and Thurnwald [23] and Asquith [3] have refuted such claims. Harvey and Thurnwald based their argument on findings from the 'Baltimore Longitudinal Study of Ageing' (BLSA), which began in 1958 and is North America's longest-running study into ageing. Harvey and Thurnwald [23] suggest that ageing does not manifest in any particular way nor follow predictable patterns. Furthermore, they state that it cannot clearly be linked to general declines in physical or mental abilities ([55], cited by Harvey and Thurnwald [23]). Similarly, Asquith [3] points out that the last few decades of improvements in health and nutrition, which have been enjoyed by those who will represent future generations of older adults, should indicate that—despite their growing number—those aged 65 years and over will present fewer health demands than previous generations of their cohort.

Far from relying on stereotypes, Harvey and Thurnwald recommend that older adults' own assessments of their lives (i.e., subjective self-ratings) should replace medically- and policy-based measurements of older persons' wellbeing and productivity. This claim is also supported in a 2011 report by leading gerontologists, Kendig and Browning [33], in which they summarise findings from research they conducted for the Australian Research Council's (ARC's) Centre of Excellence in Population Ageing Research (CEPAR). However, while Kendig and Browning assert that medical assessments of older adults' health and wellbeing do not provide accurate or meaningful results, they do not recommend that subjective self-ratings become the new standard. Instead, Kendig and Browning [33] propose that psycho-social factors be considered as means for understanding how health and wellbeing can be maintained throughout old age. They also note that while human capacities are prone to decline with age, there is much variation within the ageing process and many older adults remain capable, and productive, until very old age. In other words, older adults embody wide and varied sets of intentions and abilities, which may be mapped to a spectrum of assistive augmentation, i.e., with opportunities arising for technology to both enhance and assist with their everyday lives. Therefore, this chapter presents an overview of various issues related to ageing in place, from perspectives offered by gerontechnology, HCI (and CSCW), and interactive architecture. Given the differences that exist across these fields—for example their foci, preferred methods, tools and terminologies—our intention is to highlight key learnings that each discipline has contributed to our understanding of how technologies, and domestic environments, may be optimally designed to support ageing in place.

2 Active Ageing

The psycho-social factors proposed by Kendig and Browning [33], amongst others, align with the World Health Organization's 'Active Ageing framework' [60], which proposes that healthy ageing can be achieved by policies addressing, and individuals attending to, six key areas of life: social, physical, economic, civil, cultural and spiritual. As part of its recommendations, the Active Ageing framework acknowledges the diversity of older adults and, therefore, its incorporation of this fundamental understanding allows the framework to apply across the unique life trajectories and varied experiences of older adults. Despite this, notions of assistive augmentation have been dominated by telecare devices, which make up the majority of assistive technology solutions developed specifically for older adults. While they have helped many, they are reactive rather than preventative systems— i.e., only calling for help once something goes wrong [8]. Typically, such 'assistive technologies', whether computationally-enhanced or not, *compensate* for losses in abilities rather than maintaining or enhancing those that are important to older individuals.

In their study into older users of technology and their perceptions of the benefits to be gained from learning and using new technologies, [25] make a distinction between the *accommodation* and *assimilation* of technologies. They describe accommodation as being conducive to wellbeing and a process in which technology enhances the life of older adults through their choice to use it, i.e., not through necessity or forced adoption (in order to replace loss of physical function or cognitive ability). Assimilation, on the other hand—while also related to older adults' technology use—is presented as void of personal choice, but imbued with the stigma of 'being old' and having lost one's independence [25]. Romero et al. [52] consider such "traditional persuasive mechanisms", which monitor users and prevent accidents, as not only saving lives but also reducing the quality of lives, i.e., by impacting on one's independence and social connectedness with others.

3 Ageing in Place

Broadly speaking, ageing in place pertains to older adults ageing in the home of their choice. While for many this will mean remaining in the family home, for others it may be a home to which they have downsized or a care facility. Modifications that augment domestic environments—i.e., in order to adapt to inhabitants' changing needs and abilities, and therefore, enabling older persons to remain in their own home—support just one facet of ageing in place. Indeed, ageing in place may be viewed as a sub-stream of active ageing, which is the collective consideration of social, physical, economic, civil, cultural and spiritual aspects of ageing [60]. While this chapter focuses on life within the home, we acknowledge that ageing in place also pertains to everyday existence outside the home. Furthermore, the importance of each contributing factor of active ageing will differ according to individuals, as well as between cohorts that are delineated by a common cultural identity. For example, older Australians desire independence throughout old age whereas many Singaporean families, responding to expectations that they care for older family members, are hiring household maids to work as the primary carers for ageing family members [61]. However, from a typical Western-centric point of view, regardless of the location of one's home or the level of individual support received, e.g., from carers, friends and family, medicine and/or technology, the essential ingredients for ageing in place include autonomy and agency [51]. Consequently, ensuring accommodation can play a role in technology adoption—particularly of technologies that 'assist' or 'augment'—will likely increase the ongoing use and usefulness of the chosen technology.

Ageing in place has long been a visible research agenda within the field of gerontology, but it is far more recently that HCI has taken up its investigation. For example, Blythe and Monk [8] present a medically-skewed perspective for using technology to support an ageing population by discussing how telecare devices and smart home technologies can manage the risks of 'typical' age-related illnesses and accidents. However, Fozard [20] offers a less 'medical interventionist' view by also considering how technology might support healthy lifestyles and mitigate environmental risks—such as older persons' forced isolation or economic hardship—in order to change the course of ageing for the better. Fozard acknowledges that designing for ageing bodies means that designs cannot remain static. He writes that, much like the philosophy behind Participatory Design, design does not stop with the use of a designed solution but continues throughout its use as "the interaction between the individual aging and secular changes in the environment over time is not static".

Another prominent aspect of ageing in place pertains to the human body, and therefore, the space that it inhabits within the physical world. Not only do physical health and bodily functions determine whether autonomy and independence can be achieved, but their presence (or absence)—and level of performance—directly influence older adults' self-evaluations of their wellbeing. Likewise, the everyday activities that support active ageing, such as social engagements and spiritual pursuits, are expressed by interactions that occur between bodies, objects and space.

4   Ageing Research: Then and Now

Gerontechnology is a discipline that was borne from gerontology, which focuses on age-related issues from a social context. While gerontechnology is specifically concerned with technology for older adults, HCI's interest in technology is more general and has, since its early days, broadened to include ethnographically-inspired inquiries into interactions that involve humans and technology, as well as the behaviours that these shape and are shaped by. More formally, gerontechnology has been described as "the scientific study of aging, and technology, the development and distribution of technological products, environments, and services [for] the benefit of aging and aged people" [4]. However, during its establishment in the 1990s, it drew heavily from a "man-machine model of the relationship between humans [and] technological artefacts" ([44], p. 30), which prioritised technology development over the unique requirements of older adults, and in many examples, continues to do so. Therefore, Östlund—a Professor at the KTH Royal Institute of Technology (KTH Vetenskap och Konst), Sweden, who specialises in users' (including older adults') trust in healthcare and related technologies—offers an updated definition. According to Östlund, gerontechnology is the study of relationships between older adults and technological artefacts, which emphasises compensation (i.e., for age-related physical impairments) and accident prevention [44].

Despite criticisms that gerontechnology has devalued the social aspect of technology—i.e., only using it to locate problems in need of technical solutions [6, 43]—and of its early attempts to model the ageing process based on "man-machine-environment interaction[s]" [5], more recent work has considered the "socio-technical ensemble". For example, in a review of design methods for, and evaluations of, technology and services for older adults, Bartlett and Carroll [4] explore how and in what contexts the lives of older adults might be supported and enhanced (i.e., across areas such as housing, communication, mobility and transport, health, work, and recreation and self-fulfilment). While their review also considers how age-related losses and difficulties might be compensated for, their research acknowledges (with specific reference to modifying homes for older adults) that "significant individual differences" [4] exist within the cohort and are the results of far more than simply human-artefact interactions.

Since the 1990s HCI has also conducted research into the concepts of space and place. While predating the field of HCI, one may consider that the seeds for this research were sowed in 1965 by Sutherland's vision of a world where rooms would comprise of 'ultimate displays', i.e., computers that controlled the form and function of matter, and where the production and display of objects would occur simultaneously. Sutherland believed that not only would it be possible to display and manipulate data via the same 'interface' but also that, for example, a chair that was "displayed" would be "good enough to sit in" ([47], p. 62). Weiser [59] offered a similar vision, which spawned the research areas of ubiquitous computing, and later, tangible interaction. However, Weiser imagined a world where we would be able to interact with technology without any need for an interface, i.e., because the technology would be so embedded into everyday objects, and the fabric of our daily lives, that interfaces would essentially 'disappear' [17, 41]. Some 25 years later, Weiser's proposition endures in the potential of the Internet of Things. Similarly, Ishii and Ullmer's foundational work (1997)—into what was later called "tangible bits" [29] and 'tangible interaction'—was directly inspired by Weiser's vision, and tangible interaction has become a prominent research area, which investigates a diverse array of physical forms and modes of interaction.

5 Space: An Enabler of Tangible Interactions

While various models and frameworks have been developed to support the design and evaluation of tangible interfaces and interactions (e.g., [28, 29]), the original definition of tangible interaction developed by Ishii and Ullmer [31] described how physical objects can represent digital information and how physical actions can be mapped to the actions of computers [18]. Over the last couple of decades research into this area has grown, as has the breadth of its definition and the terms which are used to identify it. "Graspable user interfaces", "tangible user interfaces", "physical-digital interactions", and "digitally-augmented physical spaces" are the alternative labels highlighted by Hornecker and Buur [28]. However, under any of these names, the corresponding research is always concerned with removing the distinction between the representation of data and its control—opening up the possibility for tangible interactions to embody digital information as "tangible bits"; making the information directly manipulable and understood via haptic feedback.

More recent research shows that in addition to coupling the physical with the computational aspects of technology, intangible representations of data may enhance feedback by synchronously being presented, for example, via digital projections [29]. Many examples of this type of tangible interface exist, focusing on what Ishii categorises as 'Interactive Surfaces' or 'Tabletop TUIs' (ibid.), which are used for co-located collaborations. Examples of such interactive surfaces include Wellner's 'Digital Desk' [17], the 'Urban Planning Workbench' [30], the 'Envisionment and Discovery Collaboratory' [28], and the 'MR Tent' [58], amongst others. Further elaborations on the meaning of tangible interaction have been investigated by researchers such as Hornecker and Buur [28] and Fernaeus et al. [18], criticising Ullmer and Ishii's initial definition as being too data-centric. Ishii has since revised his earlier work, but in the interim, tangible interaction has evolved to include wider views of what this type of interaction can mean. Hornecker and Buur [28] introduce what they regard as "expressive-movement-centred" and "space-centred" views. The former exploits how the senses are involved in interaction with objects and how meaning is shared between users, as well as user and object, while the latter shares much in common with fields like interactive architecture and interactive art, and the way in which physical space can be used to display objects; enable social and computer interactions; and include intangible representations such as sound or video displays. Hornecker and Buur [28], and subsequently Ishii [29, 30] and Fernaeus et al. [18], are of the opinion that tangible interaction encompasses all of the above and, as such, is about much more than using physical objects to display and control digital data.

Research into sociophysical technology tends to focus individually on the discrete areas of the social and the physical. However, the basis of our interactions with the world is intrinsically part of the physicality of our bodies and the social context in which we find ourselves [7]. Within the field of HCI, Li et al. [38] evaluated features of 'quantified self' technologies—and users' practices of tracking their biometrics—to present a model for designing future systems, i.e., for use by participants' older selves. Pang et al. [46] have also recommended design tools that mediate sharing health information across distance and age gaps. Angelini et al. [2] explored how design aesthetics might encourage use.

6 Negotiating Spatial Relationships

Various studies into ageing and technology—from the fields of gerontechnology and HCI—share common aspects. However, of the two fields, it is HCI that typically acknowledges not only human users and artefacts, but also social practices, and the political and cultural contexts within which technologies are designed and used. Similarly, CSCW research is also concerned with practices and norms, specifically related to how work is negotiated amongst individuals and groups, and how mutual understanding is shared [1]. While the design of buildings is not a primary research interest of gerontechnology, HCI or CSCW, these fields are nevertheless influenced by spatial concerns, such as how space is negotiated between people and/or objects, and the relationships that exist between space, place, people and things.

With regard to ageing in place, the importance of the physical space taken up by 'the home' becomes apparent when one considers the number of everyday activities, routines and habits that occur within the walls of a domestic area, such as a family home. Additionally, as acknowledged by researchers, such as Csikszent-mihalyi and Rochberg-Halton [15] and Turkle [56], the objects with which one shares their spaces, and the memories and emotions that are embedded into them, also bear an impact on one's sense of wellbeing. Additionally, Lawson [35], a high profile architect and academic, considers architectural design as both a contributor to, and benefactor of, social and cultural influences. He also notes the importance of the individual psychological make-ups of architects and the people they design for, in design processes and outputs.

Lawson [35] illustrated how architecture, in addition to redefining landscapes and housing inhabitants, also supports and augments human behavior. For example, in front of a ticket office, the simple addition of a few poles attached to each other by rope, are enough to communicate to customers that a queue is expected to form. In fact, Lawson likens 'human behaviour in space' to a non-verbal language and suggests that space is shaped by human constructs (such as a society's rules of conduct), while also being a shaper of human behaviour. He describes the relationship between spatiality and human behaviour as adhering to rules, some of which may be "a matter of local social convention", but with many also being "a reflection of both the deep-seated needs of our psyche[s] and of the characteristics of human beings" [35]. However, very little end-user input is ever explicitly gathered, let alone included, in the design of the buildings we inhabit, and Lawson notes that many of the spaces we use are the work of professional designers who we have never met. So too in technology design, most of it has been professionally designed for us, rather than by us.

7   Interactive Architecture's Place

Interactive architecture is a relatively new area of architecture and can be found under various labels, such as 'intelligent architecture', 'responsive architecture', 'intelligent kinetic systems', 'intelligent environments', 'smart environments', 'ambient environments' and 'ambient user interfaces' [47] and 'soft space'. Interactive architecture may relate to buildings that are (i) static, but spatially and functionally adaptive, (ii) kinetic, or (iii) kinetic and 'intelligent', that is, computationally augmented by sensors, actuators and systems that manage data about inhabitants and environmental conditions (both internal and external). Sherbini and Krawczyk [54] propose that a building becomes intelligent once it includes some form of computational ability. They would categorise a building that uses computers and sensors to automatically adjust temperature and lighting as one that displays static forms of intelligence. However, others like Fox and Yeh [19] have argued that in order to be 'intelligent' a building must also move, that is, it must be kinetic.

Related to interactive architecture, Parkes et al. [47] present an extension of their work on tangible interaction and interfaces by exploring kinetic design and Kinetic Organic Interfaces (KOIs)—which fall under the broader category of Organic User Interfaces (OUIs)—to underscore the potential for tangible technology to include bi-directional capabilities, that is, the ability for physical objects to control data, and data to control physical objects. If we imagine the implications of this in a building comprised entirely of such technology, the creation of contextually-aware buildings that are capable of catering to the social and physical needs of their inhabitants seems possible. However, in the design of interactive architecture very little work has been conducted into the daily lives and needs of buildings' occupants.

Typically, work within interactive architecture focuses on structures and the technologies incorporated into their design (e.g., sensors, actuators and Heating, Ventilation and Air Conditioning (HVAC) equipment and control systems). There are of course notable exceptions, such as Croci's [14] discussion of how we relate to both urban and domestic environments, and Schnädelbach's [53] conceptual framework of adaptive architecture. Schnädelbach considers inhabitants' control of architectural features—such as can be seen in Holl's [26] 'Fukuoka Housing project' and its allowance for partitions (i.e., walls) to be moved and adapted by inhabitants, according to their needs—as well as the effect of architectural design choices on inhabitants. Also from within the field of HCI, Mäyrä et al. [39] and Leonardi et al. [37] present work that explores how spatial design impacts upon inhabitants. While Mäyrä, Soronen and Vadén focus on potential design principles for 'smart homes' Leonardi et al. explore how a group of older adults' use, and emotionally relate to, their homes. However, even though such examples consider how the design of 'intelligent' spaces impact on inhabitants, there is no universally agreed upon language they use to describe their findings or craft their design guidelines—leaving the fields of HCI and interactive architecture largely disconnected.

8 The 'User' in HCI

Within HCI, technical efforts related to ageing in place have included developing hardware that increases the accuracy of data capture [32]. Work has also been conducted into devising less obstructive applications to monitor sleep [36] and more accurate algorithms to compute gait [11]. Some human concerns, associated with using technology to support health and wellbeing, have also been explored. For example, Khovanskaya et al. [34], used critical design to propose a tracking system that actively reminds users of concerns such as controlling their privacy and the ethics of surveillance, as well as the potentially misleading interpretations of data when using such technologies. Caine et al. [10] presented the 'DigiSwitch', a device that was designed to help older adults manage privacy while monitoring their health at home. However, in their investigation of privacy concerns, Khovanskaya et al. [34] did not delve into how these concerns were experienced by older users of their tracking system. Another point of difference between the work of Caine et al. [9], Caine et al. [10] and Khovanskaya et al. [34] is that the Caine et al. studies focused on the device as a technical achievement, that is, part of an act of agency by older users rather than an exploration of their privacy concerns.

User-centred design is often leveraged within HCI research, in order to focus the design of technology upon intended users (i.e., their abilities, desires and needs, amongst other considerations) and likely contexts of use [27, 42]. However, within HCI, there remains a general lack of deep and sustained engagement, and understanding, of people's use of technologies that support wellbeing and active ageing. Despite some human concerns being raised or encouraged (e.g., [34]), there have not been any deep explorations into how these concerns are experienced by users or how they affect people's use of the technology. However, a Participatory Design workshop to co-design possible mobile health applications—targeted for use by older adults [16]—aimed at giving older adults a 'voice' and ensuring that their needs were met. However, while Davidson and Jensen's study produced novel ideas, to date, these have not yet progressed from being concepts. Furthermore, their study was function-led, that is, focused on what participants would like to track without investigating how such applications might actually be used and experienced in everyday life.

9 Innovating Usability

It is quite common for technology design research, particularly work that adopts a medically-oriented view of wellbeing, to produce tools that track, and provide feedback on, the health and physical wellbeing of the 'average person'. Consequently, such technologies are unlikely to be suitable for older adults whose bodies, and particular age-related requirements, are exponentially diverse and unique [50] compared to younger users. This suggests a need for more tailored approaches to designing technologies aimed at supporting the wellbeing of older people, and their desires to age in place. Additionally, resultant technologies would benefit—in both user adoption numbers and continued use—if they could avoid the stigma of being labelled 'solutions for old people' while offering flexibility of use, which could support users' changing needs due to the ageing process, i.e., by allowing users to easily adapt and modify technologies.

An example of such work is offered by Gaver et al. [21] who utilised design-driven innovation in a research project called 'Presence'. By using cultural probes to explore the desires and pleasures of older users, and beginning with the users themselves to design technology—not to solve problems but to provide opportunities—Gaver et al. turned traditional design on its head. The cultural probes were used as sources for inspiration and insight into what aspects of life were important to the older participants, and provided views that would unlikely have been uncovered by more conventional means. Additionally, given the Presence project involved three groups of older adults that differed in geography, culture and social structure, the cultural probes allowed for the nuances of these differences to appear rather than risk falling into a one-size-fits-all solution, which can still be seen in gerontechnology and HCI work that is targeted toward older adults.

10 Conclusion

While interactive architecture readily expands in the advancement of materials for intelligent kinetic structures the research from this field typically ignores the concerns of older adults and ageing in place. HCI has demonstrated that usability alone is not enough to encourage the use of technology by older adults—although functionality needs to be understood and confidence in its use and usefulness is necessary for continued adoption—and research from within gerontechnology suggests that a focus on end-users' accommodation is more conducive to technology adoption than assimilation [25].

Likewise, by considering how a home could be considered as a participant in a social exchange, that is, a companion who can respond to sociophysical interactions, and to whom emotional and psychological attachment is formed—the design of a contextually aware room, smart house or intelligent kinetic building becomes something much more than a technical endeavour. Instead, designers can begin by considering the 'opportunities' and 'interesting situations' that older adults associate with ideas of home and ageing in place. Similarly, considering the 'cost' versus the 'benefit' of adopting new technology has much to do with understanding where on the continuum of use it falls for particular users. Assistive augmentation may include assistive technologies and be medically-skewed, however, as the theory that dispels stereotypes about older adults suggests, it is also important that it can also support and enhance the abilities, and activities, that older adults already enjoy.

Further research into designing domestic environments for ageing well—from across the fields of gerontechnology, HCI and interactive architecture—would do well to consider the everyday concerns of potential users, in addition to their understanding about, and feelings toward, future technologies. Participatory Design methods could provide designers with insights into such areas, and design-driven innovation—e.g., accessed through the use of cultural probes—would encourage yet-to-be-realised ideas to emerge. In order for such learnings to benefit the multi-disciplinary teams required for the creation of domestic environments that support ageing in place, a shared set of methods and design considerations are necessary; however, as yet, no such 'common language' exists.

References

1. Ackerman MS, Starr B (1995) Social activity indicators: interface components for CSCW systems. In: Paper presented to the proceedings of the 8th annual ACM symposium on user interface and software technology, Pittsburgh, Pennsylvania, USA
2. Angelini L, Caon M, Carrino S, Bergeron L, Nyffeler N, Jean-Mairet M, Mugellini E (2013) Designing a desirable smart bracelet for older adults. In: International joint conference on pervasive and ubiquitous computing (UbiComp' 13), Zurich, Switzerland, pp 425–434
3. Asquith N (2009) Positive ageing, neoliberalism and Australian sociology. J Soc 45(3):255–269
4. Bartlett H, Carroll M (2010) Capacity building in ageing research: key successes and challenges of the Australian experience. Gener Rev 20(4):1–4
5. Baskinger M (2007a) Autonomy + the aging population: designing empowerment into home appliances. In: Feijs L, Kyffin S, Young B (eds) DeSForM Design and semantics of form and movement, pp 133–146
6. Baskinger M (2007b) Experientializing home appliances to empower the aging population for autonomous living. In: Proceedings of the 2007 conference on designing for user eXperiences. ACM, p 14
7. Bekker T, Sturm J, Barakova E (2010) Design for social interaction through physical play in diverse contexts of use. Pers Ubiquit Comput 14(5):381–383
8. Blythe M, Monk A (2005) Net Neighbours: adapting HCI methods to cross the digital divide. Interact Comput 17(1):35–56
9. Caine KE, Zimmerman CY, Schall-Zimmerman Z, Hazlewood WR, Camp LJ, Connelly KH, Huber LL, Shankar K (2011) DigiSwitch: a device to allow older adults to monitor and direct the collection and transmission of health information collected at home. J Med Syst, pp 1181–1195
10. Caine KE, Zimmerman CY, Schall-Zimmerman Z, Hazlewood WR, Sulgrove AC, Camp LJ, Connelly KH, Huber LL, Shankar K (2010) DigiSwitch: design and evaluation of a device for older adults to preserve privacy while monitoring health at home. In: International health informatics (IHI '10). Arlington, Virginia, USA, pp 153–162
11. Cheng Q, Juen J, Li Y, Prieto-Centurion V, Krishnan JA, Schatz BR (2013) GaitTrack: Health monitoring of body motion from spatio-temporal parameters of simple smartphones. In: Bioinformatics, computational biology and biomedical informatics
12. Cornejo R, Hernández D, Favela J, Tentori M, Ochoa S (2012) Persuading older adults to socialize and exercise through ambient games. In: 2012 6th International Conference on, Pervasive Computing Technologies for Healthcare (Pervasive Health), IEEE, pp 215–218
13. Cornejo R, Tentori M, Favela J (2013) Ambient awareness to strengthen the family social network of older adults. In: Computer supported cooperative work (CSCW), pp 1–36

14. Croci V (2010) Relational interactive architecture. Archit Des 80(3):122–125
15. Csikszentmihalyi M, Rochberg-Halton E (1981) The meaning of things: domestic symbols and the self. Cambridge University Press, Cambridge, Massachusetts
16. Davidson JL, Jensen C (2013) What health topics older adults want to track: a participatory design study. In: International conference on computers and accessibility (ASSETS '13). Bellevue, WA, USA, pp 1–8
17. Dourish P (2001) Where the action is: the foundations of embodied interaction, Kindle edn. MIT Press, Cambridge, Massachusetts
18. Fernaeus Y, Tholander J, Jonsson M (2008) Beyond representations: towards an action-centric perspective on tangible interaction. Int J Arts Technol 1(3/4):249–267
19. Fox MA, Yeh BP (1999) Intelligent kinetic systems. Prep MANSEE 99
20. Fozard JL (2002) Gerontechnology—beyond ergonomics and universal design. Gerontechnol 1(3):137–139
21. Gaver WH, Hooker B, Dunne A, Farrington P (2001) The presence project. RCA Computer Related Design Research, London
22. Grosinger J, Vetere F, Fitzpatrick G (2012) Agile life: addressing knowledge and social motivations for active aging. In: Paper presented to the OZCHI'12, Melbourne, Victoria, Australia
23. Harvey PW, Thurnwald I (2009) Ageing well, ageing productively: the essential contribution of Australia's ageing population to the social and economic prosperity of the nation. Health Soc Rev 18(4):379–386
24. Hazan H (1994) Old age: constructions and deconstructions, Cambridge University Press
25. Hernández-Encuentra E, Pousada M, Gómez-Zúñiga B (2009) ICT and older people: beyond usability. Educ Gerontol 35(3):226–245
26. Holl S (2010) Housing Complex, Fukuoka. In: Ebner P, Herrmann E, Höllbacher R, Kuntscher M, Wietzorrek U (eds) Typology +. Berhäuser Verlag AG, Berlin
27. Holtzblatt K, Beyer H (2014) Contextual design: evolved, Morgan & Claypool
28. Hornecker E, Buur J (2006) Getting a grip on tangible interaction: a framework on physical space and social interaction. In: Paper presented to the proceedings of the SIGCHI conference on human factors in computing systems, Montréal, Québec, Canada, pp 22–27, April, 2006
29. Ishii H (2008a) Tangible bits: beyond pixels. In: Paper presented to the proceedings of the 2nd international conference on tangible and embedded interaction, Bonn, Germany
30. Ishii H (2008) The tangible user interface and its evolution. Commun ACM 51(6):32–36
31. Ishii H, Ullmer B (1997) Tangible bits: towards seamless interfaces between people, bits and atoms. In: Proceedings of the ACM SIGCHI conference on human factors in computing systems, ACM, pp 234–241
32. Jalaliniya S, Pederson T (2012) A wearable kid's health monitoring system on smartphone.In: NordiCHI, pp 791–792
33. Kendig H, Browning C (2011) Directions for ageing well in a healthy Australia. Acad Soc Sci 31(2):23–30
34. Khovanskaya V, Baumer EP, Cosley D, Voida S, Gay G (2013) Everybody knows what you're doing: a critical design approach to personal informatics. In: CHI, pp 3403–3412
35. Lawson B (2001), The language of space, Architectural Press
36. Lawson S, Jamison-Powell S, Garbett A, Linehan C, Kucharczyk E, Verbaan S, Rowland DA, Morgan K (2013) Validating a mobile phone application for the everyday, unobtrusive, objective measurement of sleep. In: Paper presented to the proceedings of the SIGCHI conference on human factors in computing systems, Paris, France
37. Leonardi C, Mennecozzi C, Not E, Pianesi F, Zancanaro M, Gennai F, Cristoforetti A (2009) Knocking on elders' door: investigating the functional and emotional geography of their domestic space. In: Paper presented to the CHI 2009, Boston, MA, USA
38. Li I. Dey AK, Forlizzi J (2011) Understanding my data, myself: supporting self-reflection with ubicomp technologies. UbiComp, pp 405–414
39. Mäyrä F, Soronen A, Vadén T (2004) Future proactive homes: some design principles. NordiCHI, Tampere, Finland

40. National Health and Medical Research Council (2004) Ageing well, ageing productively, Commonwealth of Australia. http://www.nhmrc.gov.au/grants/types-funding/-z-list-funding-types/ageing-well-ageing-productively-program-grant
41. Negroponte N (2009) Being digital. Random House, Audible.com edn
42. Norman DA, Draper SW (1986) User-centred system design: new perspectives on HCI. Lawrence Erlbaum Associates Inc, Hillsdale, New Jersey
43. Östlund B (2004) Social science research on technology and the elderly—does it exist? Sci Stud 17(2):44–62
44. Östlund B (2005) Design paradigms and misunderstood technology: the case of older users. In: Östlund B (ed) Young technologies in old hands: an international view on senior citizens' utilization of ICT. DJOF Publishing, Copenhagen, pp 25–39
45. Östlund B (2008) The revival of research circles: meeting the needs of modern aging and the third age. Educ Gerontol 34(4):255–266
46. Pang CE, Neustaedter C, Riecke BE, Oduor E, Hillman S (2013) Technology preferences and routines for sharing health information during the treatment of a chronic illness. In: CHI, pp 1759–1768
47. Parkes A, Poupyrev I, Ishii H (2008) Designing kinetic interactions for organic user interfaces. Commun ACM 51(6):58–65
48. Pedell S, Vetere F, Kulik L, Ozanne E, Gruner A (2010) Social isolation of older people: the role of domestic technologies. In: Paper presented to the OzCHI 2010 CHISIG, ACM, Brisbane, Australia
49. Riche Y, Mackay W (2010) PeerCare: supporting awareness of rhythms and routines for better aging in place. Comput Supported Coop Work (CSCW) 19(1):73–104
50. Robertson T, Durick J, Brereton M, Vaisutis K, Vetere F, Nansen B, Howard S (2013) Emerging technologies and the contextual and contingent experiences of ageing well. In: Kotzé P, Marsden G, Lindgaard G, Wesson J, Winckler M (eds) Human-computer interaction —INTERACT 2013, vol 8119. Springer, Berlin Heidelberg, pp 582–589
51. Robertson T, Durick J, Brereton M, Vetere F, Howard S, Nansen B (2012) Knowing our users: scoping interviews in design research with ageing participants. In: Proceedings of the 24th Australian computer-human interaction conference, ACM, pp 517–520
52. Romero N, Sturm J, Bekker T, de Valk L, Kruitwagen S (2010) Playful persuasion to support older adults' social and physical activities. Interact Comput 22(6):485–495
53. Schnädelbach H (2010) Adaptive architecture–a conceptual framework. In: Interaction of architecture, media and social phenomena, p 523
54. Sherbini K, Krawczyk R (2004) Overview of intelligent architecture. In: Paper presented to the 1st ASCAAD international conference, e-design in architecture, KFUPM, Dhahran, Saudi Arabia, December, 2004
55. Sperry L, Prosen H (1996) Aging in the twenty-first century: a developmental perspective. Garland, New York
56. Turkle S (ed) (2008) The inner history of devices. The MIT Press, Cambridge, Massachusetts
57. Vaisutis K, Brereton M, Robertson T, Vetere F, Durick J, Nansen B, Buys L (2014) Invisible connections: investigating older people's emotions and social relations around objects. In: Paper presented to the CHI 2014, Toronto, Cananda
58. Wagner I (2011) Building urban narratives: collaborative site-seeing and envisioning in the MR tent. Comput Supported Coop Work (CSCW) 21(1):1–42
59. Weiser M (1991) The computer for the 21st century. Sci Am 265(3):94–104
60. World Health Organization (WHO) 2002 Active ageing: a policy framework, pp 1–59
61. Yeoh BSA, Huang S (2009) Foreign domestic workers and home-based care for elders in Singapore. J Aging Soc Policy 22(1):69–88

Sensory Conversation: An Interactive Environment to Augment Social Communication in Autistic Children

Scott Andrew Brown and Petra Gemeinboeck

1 Introduction

For autistic people, sensory interactions throughout daily life are augmented by profound and impactful experiences, not consciously considered by many in the mainstream, or *neurotypical* population. The project outlined in this chapter views these interactions as important elements of a communicative ontology, and discusses a responsive sensory environment that has been designed to engage sensory expression and motivate social communication between autistic children[1] and their parents. Constructed as a dome tent-like structure, this material artefact follows a history of using multisensory environments in therapeutic settings as a place of relaxation and play. In the context of this study, the responsive dome environment (RDE) developed and tested the design method of *conversational probes*. Specifically, this research investigates ways to facilitate and reflect on emergent conversational relationships between children, their parent and the researcher and, in particular, aims to support unique expressions and shared experiences through embodied interaction with the sensory environment.

The design process of this responsive sensory environment involved iterative observational studies focused on a group of autistic and neurotypical children and their parents in a space for non-directed and play-like expression. Viewed as active

---

[1]I acknowledge the common usage of person-first language in most academic writing relating to autistic persons. However, the identity-first language used in this chapter is done so out of respect for autism self-advocates, who wish to recognise autism as an important part of their being in the world.

S.A. Brown (✉) · P. Gemeinboeck
Creative Robotics Lab, UNSW Sydney | Art & Design, Paddington, NSW, Australia
e-mail: scott.brown@unsw.edu.au

P. Gemeinboeck
e-mail: petra@unsw.edu.au

© Springer Nature Singapore Pte Ltd. 2018
J. Huber et al. (eds.), *Assistive Augmentation*, Cognitive Science
and Technology, https://doi.org/10.1007/978-981-10-6404-3_8

partners in the research and design process, the children and parents were invited to explore a tangible system, which was designed to explore the relationship between interactive experiences and social communication between participants. The genesis for this research is a case study undertaken by the first author, working with a 2 year old autistic girl in 2011 [9]. This earlier project examined the potential for embodied interactive devices to facilitate aesthetic agency and found serendipitous success with a glance between mother and child. The child displayed an awareness of her agency in generating sensory feedback from the device, and in turn she would turn her attention to her mother, checking that she was sharing the experience and echoing her communication with the responsive object. Recognised as 'joint attention' by an occupational therapist present, this seemingly simple act was an indicator of social interaction; an area of diagnosed deficit for this child. Reflection of agency afforded by the interactive device precipitated social interaction, initiated by the child toward her mother. This unexpected event led to the research described in this chapter attempting to further understand how interactive experiences can facilitate or augment social conversation.

One of the aims of the responsive dome environment (RDE) is to engage a child's agency by reflecting physical presence through embodied interaction. The importance of embodiment can be appreciated through the definition put forward by Paul Dourish: "Embodied Interaction is the creation, manipulation, and sharing of meaning through engaged interaction with artifacts" [12, p. 126]. The premise being that this mode of interaction provides each child with an opportunity to initiate a conversation of their own making, not only with the responsive system, but with others that share the space with them. The RDE allows children to initiate interaction on their own terms, keeping the research process receptive to "what children want to do as opposed to what adults expect of them" [13] and mitigating inequitable power relationships between participants and researchers.

Aligned with the stated aim of participant self-direction during interaction with the dome environment, is the notion that neurotypical populations need to develop awareness for alternative modes of communication and an appreciation for *neurodiversity*, a term brought into mainstream discussion by Steve Silberman's book, *Neurotribes: The Legacy of Autism and How to Think Smarter About People Who Think Differently* (2015). The premise also received recognition through autism activists such as Amanda Baggs, whose widely-viewed video *In My Language* [4] points to a neurotypical preference for language as the primary mode of expression. In her video, Baggs shows herself "in constant communication with every aspect of her environment"—with and including her own body—which she embraces as her native language, rather than considering the actions as stereotypical 'stimming' behaviour, often associated with autism. Although the language Baggs presents is not intended as a learnable semiotic system, she is engaged in an embodied sensory conversation, which neurotypical people are equally capable, even if not practiced. By providing a controlled and controllable stage for sensory conversations to take place, a key goal of the RDE is to facilitate shared experiences which are conducive to social communication.

The conversations discussed here are led by an interaction design approach, rather than that of scientific rationalism. Therefore experiential affect is viewed in an 'interactional' rather than 'informational' context, and as presented by Boehner et al., "we do not try to simplify complexity but instead to augment it" [5, p. 66]. In other words, experiences are recognised as being unique, emergent and dynamic, rather than discrete and generalisable. This is pertinent when working with an autistic population, where the phrase, "when you've met one person with autism, you've met one person with autism" describes the spectrum nature of the condition.

1.1 Sensory Processing and Autism

The term 'autism' is derived from the Greek 'auto', meaning 'self' and "...proclaims the apparent mental involution or self-absorption of autistic people", including a difficulty engaging in social interactions [45]. As a spectrum condition, autism by its very nature is complex and unique in every individual with a diagnosis. While there are diagnostic criteria for autism which address a "Hyper- or hyporeactivity to sensory input or unusual interests in sensory aspects of the environment" (2013), there is no strict definition on the type or amount of sensory processing issues that someone with an autism might be faced with.

There has been a recent shift in the way that these sensory sensitivities found in autism are perceived: rather than being seen as an impairment, it is accepted as a part of an autism self-advocacy movement that uses the more inclusive term of 'neurodiversity' [11, p. 2296], with each person reacting "to stimuli in very idiosyncratic ways" (Iarocci and McDonald [23 p. 79]. Accounts of "atypical reactions to sensory stimuli" and "unusual attention to parts rather than wholes" [23, p. 77] can be traced back to Leo Kanner, whose work formed the early basis of the study of autism.[2] The impact of atypical sensory processing has an impact on developmental milestones, potentially with "downstream effects" on perceptual engagement [23, p. 77]. To explore this progression, this research project worked with children of developmental age (3–6 years). Iarocci and McDonald believe that perceptual development is built upon the ability to first focus on a singular sensory experience, but then "integrate multiple sources of input" [23, p. 81].

With the general environment a potentially overwhelming sensory space, the aforementioned tendency for those with autism to focus on a singular experience suggests that controllable sensory environments present an interesting opportunity to examine these questions in a systematic way. Being neurologically agnostic, interactive systems are well placed to augment expression and support observation of sensory experiences. The control and reliability of a programmed responsive

---

[2]As discussed in Steve Silberman's book, *Neurotribes: The Legacy of Autism and How to Think Smarter About People Who Think Differently* (2015), Leo Kanner's work is now considered controversial in the way that it addresses autistic people. However, the acknowledgement of autism as a specific condition was almost non-existent before Kanner's research in the 1940s.

system lends itself to reducing stress in people who have difficulty in regulating their sensory inputs. Deborah Lipsky, a high-functioning autistic individual, puts it thus: "I can't emphasize enough how critical it is to understand that staying on a script is the sole means of keeping anxiety at a minimum. Even the smallest breach becomes a crisis because all we register at that moment is unpredictability. We fear unpredictability above all else because we are out of control of our environment" (Lipsky [24] cited in [39]).

This presents both an opportunity and a challenge to the design researcher: as a study involving a programmed system, autistic children are able to interact with the space and expect a regular and reliable response from the programmed system, consequently avoiding the anxiety that Lipsky describes. Indeed, much of the research which looks at technology-based interaction for autistic people have used this to great effect [26, 30, 35, 41]. However, the reliability of a simple programmed system can quickly result in supporting the repetitive behaviours that are symptomatic of an attempt to regulate overwhelming sensory input in an autistic person, yet can make social engagement difficult. The RDE attempts to reconcile these behaviours by looking to the work of cybernetician Gordon Pask, in particular his conversational performative system, *Musicolor*, explored in more detail later in this chapter.

1.2 Multisensory Environments

Multisensory experiences with a physical environment support learning in a constructionist sense of the word: it is embodied, situated, self-directed and engaged through conversations with artefacts [1]. Wright and McCarthy likewise frame experience through a holistic lens, "that treats as inseparable, people's intellectual, sensual, and emotional responses, and that conceptualizes self, artefacts, and settings as multiple centres of value interacting with each other" [42, p. 638]. In addition, this approach to experience supports our decision to not employ screen-based technologies in exploring sensory augmentation.

The responsive dome environment (RDE) fits within a history of multisensory environments (MSEs) being used in a therapeutic context. MSE is a catch-all term for a variety of spaces that present multiple modes of sensory engagement or diversion. In therapeutic usage, these spaces have a lineage that began with the Snoezelen® room in the 1970s. A conjunct of two Dutch words: *snuffelen* and *doezelen*, the term Snoezelen® describes exploration in a relaxed state [6, p. 139] and this remains a focus for the RDE. Snoezelen® rooms were originally created by two therapists working with people with developmental disabilities [28, p. 305], however their efficacy in achieving therapeutic or educational outcomes remains controversial and MSEs are currently used primarily as a diversional tool to support other interventions or tasks. Sensory experiences in these spaces are generally passive, rather than engaging modes of interaction that move beyond a *reactive* or *fixed* response [16, p. 70].

There are limited examples of using *dynamic interaction*—that is, "the precise way that 'input affects output' can itself change" [16, p. 70] to facilitate or augment sensory experiences in multisensory environments. One of these is the MEDIATE environment [32], which uses camera tracking techniques to allow autistic children embodied interactions with a visually responsive space. Similarly to the RDE, one of the goals of MEDIATE is for "...the children to have the chance to play, explore and be creative in a predictable, controllable and safe space" [32, p. 110], which positions the project in an emergent and experimental, rather than therapeutic context. Providing children the opportunity to become aware of their agency or control through "developmentally appropriate types of play" [27, pp. 174–175], the approach of both MEDIATE and the RDE aims to introduce awareness of agency through embodied sensory play-like interactions, which does not rely on language: "Sensation happens before we've given word or thought to what is sensed, before we make sense of it" [40].

2 Spatial and Study Design

The responsive dome environment (RDE) is a dome-shaped tent structure with a floor diameter of 3 metres and constructed from a translucent acrylic textile (Fig. 1). This material was chosen to diffuse coloured light projected from a 15 head lighting system that surrounds the structure, along with a quadrophonic sound system.

Fig. 1 Responsive dome environment with first iteration of table interface

The RDE is designed to be modular, mobile and fit within standard Australian classroom dimensions.

The outer membrane of the RDE can be thought of as the context for which there is a "shared language" [31] between the participant(s) and the feedback system. Using physical interaction with a tactile interface placed in the centre of the internal space, the architectural structure reflects the engagement through mapped audio-visual responses. While we will discuss the behaviour of this system in the next section, we will note here that this structural surface is also a region of conversation. It is the internal space of the RDE where participant(s) are afforded the agency and control to engage with the dynamic nature of the system through changing light and sound across a physical membrane which elicits interpretation and conversation.

The dialogue which emerges from experiences in the space is not only a product of whether the child has an autism diagnosis, but is a reflection of their sociocultural background; each participant brings their own sensory language to the space and will experience conversation in unique ways. One of the major design goals of the RDE was to augment and embody experiences for children and their parents within a shared space. In this sense, the approach is that of participatory design which Wright and McCarthy identify as requiring a conversational foundation between researcher(s) and participant(s): "how do you attend to the phenomenon and people, engage with the person's experience… At one level, the answer is in the dialogue. It is in the listening, the responsiveness, and the openness of dialogue" [43, p. 106].

Designing a space for dialogue or conversation with a participant population of 3–6 year old children (regardless of autism diagnosis) presents a range of communication challenges. In designing technologies for children, Allison Druin points out that young children may struggle to express their thoughts, often resulting in researchers instead asking an adult to speak on their behalf. Rather than a child verbalising feedback, "much of what they say may be in their actions and therefore, needs to be interpreted within the context of concrete experiences" [14, pp. 1–2]. With this in mind, our approach to opening up conversation with children was that of a combining observational case studies to reveal participant narratives with post-study interviews to "clarify children's motivations and pinpoint specific reactions to particular content" [14, p. 6]. The study included three iterations of observational studies, with a participant population of four neurotypical children, with the final two iterations also undertaken by two children with an autism diagnosis. All of the children were accompanied by their mother while in the RDE for a self-determined period of time, followed immediately by a semi-structured interview which reviewed video footage of the interactive experience to guide discussion.

The combination of observational case studies and interviews construct a story of experience, which is developed through a dialogue with participants [43, p. 55]. While the conversational nature of an interview is obvious, the ability for an interactive experience to constitute a similar exchange is less so. Wright and McCarthy suggest that participant use of a technology addresses the designers assumptions and as such, "a technology is both a theory of and a hypothesis about use. It is a question put to the user by the designer" [43, p. 22]. This is the first step in a conversational exchange, where the ensuing dialogue is situated in the experience of the technology, finally being analysed by the researcher.

Design challenges addressed in the construction of the RDE (pre-participation), were identified based on the literature around sensory processing difficulties associated with autism, as well as site visits to autism schools. For example, the scale of the RDE is such that it is perceived as being contained (and separate from the potentially anxiety provoking space of the research lab or school classroom), but not so small as to be claustrophobic. The structural design supports an experience of the RDE as a milieu of its own, and the external positioning of lights and speakers creates a muted, calming display, despite responding dynamically, to avoid sensory hyper-stimulation. The visual and auditory feedback itself has not been designed to be representational, that is, it does not attempt to identify or generalise experience through the use of specific colours or sounds, rather, it is designed to facilitate emergent interpretation and expressions.

Based on a low, hexagonal table form (Fig. 1), the design development involved three iterative stages in line with the three aforementioned conversational studies to open-up the design goals for the next iteration. Each tactile interface situated parent and child in a conversational relationship with each other, much like sitting at a round table. While the object on the surface of this table changed between study iterations based on the feedback from participants (coloured blocks; wooden pegs; and coloured buttons respectively), each was based on a form which encouraged play and collaboration between parent and child, continuing the design effort to augment an embodied conversational experience.

3   Interaction and Conversation Design

Interactivity is a term commonly applied to describe digital technology, yet is very broad and under-defined in its use [25, p. 85]. The perspective taken toward the activity of interaction when designing the RDE looks to methods of exploring personal experience, over an approach which favours use of technology as an end in itself. One framing can be found in interactive art, where a goal-oriented focus is sidestepped and rather than "a goal or exit, a reward in the conventional sense" [38] participants are engaged in a journey or negotiation-based, meaningful experience. It is the *experience* or *conversation* which is the primary consideration in many artistic interactive encounters. For designers Dubberly et al. too, interaction is not simply the functional aspect of an artefact, but is seen as "a way of framing the relationship between people and objects designed for them—and thus a way of framing the activity of design" [16, p. 69]. This chapter posits that interaction can also facilitate the relationship between participants and designers within practices of action research. Throughout this process, there is an ongoing conversation that takes place between these stakeholders, even though it may not be a (seemingly) direct one. By presenting participants with a open-ended approach to interpreting interaction with objects, emergent experiences form a multi-faceted dialogue.

Similarly to Simon Penny's thoughts on the aesthetics of interaction, regions of interpretation—and therefore conversation—are spread across three design concerns of the RDE: "the material artefact, the code/machine system, and the dynamics of interaction" [34, p. 99]. While the previous section addressed the RDE as a "material artefact", the tangible interface inside the RDE is framed differently; it is not only closely aligned with what Penny describes as the "code/machine system", but also links to his concept of situated *performativity* in interaction, where "the doing of the action by the subject in the context of the work is what constitutes the experience of the work. It is less the destination, or chain of destinations, and more the temporal process which constitutes the experience" [34, p. 83]. Like the approach taken in this research, Penny sees interactivity not as an act of technology-focused instrumentation, but rather as a situated and embodied performance, where meaning-creation takes place.

Locating this performance in the centre of the RDE allows the participants to experience their agency by being surrounded by the feedback of the space. By controlling a tangible object (blocks, pegs or buttons, depending on the design iteration), the mapped output of the audiovisual system will respond to augment that sensory experience: pressing a red button results in being surrounded by red light and a musical note playing from the mapped location of the interaction; pressing the adjacent green button results in green light and a successive note; and so on.

It's important to note here that the interaction supported by the RDE is not designed to contrast autistic children with neurotypical children. That is, the skills required for interacting are of a basic fine motor rather than cognitive focus. This embodied approach to sensory interaction provides an experience which can be shared cooperatively between parent and child, regardless of the child's diagnosis.

The tactile interface became the location of engagement, and was developed through three design iterations, with the style of interaction being adapted based on observational feedback and interview responses. The first (Fig. 1) and second iterations were built on camera-tracking systems, which used the position of objects (coloured blocks and wooden pegs respectively) to trigger light and sound feedback, mapped to corresponding positions on the RDE surface. The challenge this aimed to address was how to make the child immediately aware of their ability or agency for embodied control of the RDE, without explicit instruction. These designs used what Preece et al. describe as a 'manipulating' style of interaction, the design assumption being that this tangible approach would leverage the child's experience of play and "familiar knowledge of how to interact with objects" [37, p. 48]. Study results revealed that while the tactile and play-like interaction was familiar to the participants, the potential of these artefacts to control the RDE confused them—their relationship to the audiovisual feedback system was ambiguous, and ultimately the complexity of the interaction resulted in no shared language being established between participant and system.

Concepts of *interactivity*, *embodiment* and *conversation* are intertwined in this research and as such, the participant(s) must be aware of their interactive agency to engage in embodied conversations. Put differently, Penny states: "the system must present effects which are perceived by the user as being related to their actions. Without this there is no perception of interactivity" [34, p. 80]. In the RDE feedback system without clearly prescribed tasks or goals, the ambiguity of the first interface iterations led to a breakdown in communication between the participant and system. While conversation was the desired aim during the studies, it became apparent that a breakdown can provide a focus for discussion during post-study interviews.

The feedback from the interviews identified that problems with interaction were less an issue of goal or task-finding and more related to difficulty in mapping feedback between interface and RDE. The response from the children showed that they were conscious of a relationship between the colour of the object of control (either block or wooden peg), but were less certain on how (if at all) position on the table surface mapped to lighting and sound on the RDE. One of the autistic participants pointed out their attraction to colour as more interesting than understanding a spatial pattern in the system:

child: *Also I found out that these blocks were actually making those lights.*
interviewer: *Oh, you noticed that they were making the lights!*
child: *Yeah.*
interviewer: *Great. Do you have a favourite colour out of the blocks?*
child: *(holding yellow block)*
interviewer: *Yellow?*
child: *Yeah, that's why I picked this one.*

Both the researcher and the child's parent were surprised by this revelation during the interview, both having made the earlier assumption that his behaviour in the RDE displayed little awareness of control in the space (the child held his gaze on the interface alone). His reflection not only supports the use of an open approach to interviewing participants post-study, but builds on Dourish's proposal that conversations are embodied and "denote a form of participative status" [12, p. 101]. As an example to support the inclusion of the child as a unique centre of knowledge, this shows that given the freedom and empowerment to become a part of the conversation, the child's contribution leads to unexpected insights and challenges assumptions.

3.1 Cybernetic Conversation

It is important to note that the definition of *conversation* used in the research described here is derived from cybernetic systems, in particular the work of Gordon Pask. Here, conversational systems are cybernetic, in that they are goal-oriented systems. As research that is situated in the field of interaction design, it is certainly worth looking at the structure of these systems, to examine how conversation might be best facilitated by the designer.

Cybernetics is an important reference to this research, as it recognises systems that exist in a biological context, not simply the digital or mechanical world and indeed there are systems connecting both. The study of these systems "is concerned not so much with what systems consist of, but how they function" [21, p. 2]. For Ross Ashby, an important pioneer of cybernetic systems, difference was the most fundamental concept in cybernetics: "either that two things are recognisably different or that one thing has changed with time" [3, p. 9]. Examining difference is used not only to compare different systems, but also different states within the same system; the point at which comparison takes place is the *feedback loop*. This loop— or dialogue between systems—is also the conversation, where Gordon Pask believed learning between systems can take place. Action, feedback and adjustment is a looping cybernetic process that allows participants, involved in the conversational loop, to not only become aware of their ability to control a system, but also to reflect on their relationship with the system and their role in this embodied conversation.

The relevance of a cybernetic perspective to action research—certainly participatory design—is made clear by Andrew Pickering, when comparing it against design which takes an absolute attitude to artefact production: "the cybernetic approach entailed instead a continuing interaction with materials, human and nonhuman, to explore what might be achieved—what one might call an *evolutionary* approach to design, that necessarily entailed a degree of *respect* for the other" [36, p. 32]. There may be no better example of a cybernetic machine which displays this sense of exploration and respect than Gordon Pask's *Musicolour* environment, produced 1950s and 60s.

In an effort to design an environment that augments a sensory ontology, the programmed behaviour of the RDE references the *Musicolour* system of Pask which was "inspired by the concept of synaesthesia and the general proposition that the aesthetic value of a work can be enhanced if the work is simultaneously presented in more than one sensory modality" [33, p. 77]. The conversation in *Musicolour* took place between the input of musicians and feedback of lighting, where the environment would respond to tempo, key, rhythm and other elements of the musical performance. However, at some point in the performance, the relationship would change: the system would become *bored* by a lack of novelty in the music, no longer mirroring the rhythm or tempo of the musicians, but creating non-mapped lighting patterns instead. At this point, the performer could choose to respond in turn to the new patterns of *Musicolour*, feeding back as the lighting became the input to the system. For Pask, this changeability in cybernetic goal-seeking formed part of his Conversation Theory and introduced the concept of *choice*, which Pask saw as an important tenet of learning through conversation [15].

Pask's *Musicolour* also provided an inspirational example of how participants project behavioural characteristics onto the system, which at times "gets bored… given a repetitive input, the system 'directs its attention' to the potentially novel"

[33, p. 80]. When a system displays a behaviour such as this (real or perceived), it can motivate the participant to respond with changing or emerging behaviours. Presented through a cybernetic lens, the output of one system causes the observing system to readjust their own goal(s).

3.2 Interface and Control

The tools and approach described in this chapter have been employed to facilitate emergent conversations initiated by participants. In the RDE study, this has been explored through the lens of sensory-based, rather than linguistically-reliant conversation, motivated by the ability (or lack thereof) for embodied control of environmental lighting and sound feedback. The difficulty described in the first two interface iterations to facilitate conversation was identified as being linked to ambiguity of control, which led to confusion and disinterest. As Druin points out, children can become bored and distracted if they are not engaged in a task or activity [14, p. 1], so if attention is not captured quickly, it will likely be difficult to establish or augment conversation.

The final interface design (Fig. 2) was designed to combat distraction by being as explicit as possible in mapping and control of the RDE. This was also an attempt to clearly establish a common language between participant and system, thus allowing conversation to be explored in increasingly complex ways. As Simon

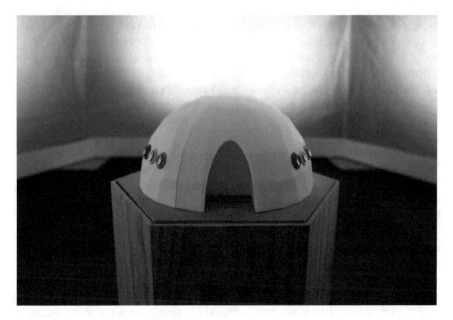

Fig. 2 Third iteration of table interface

Penny points out, "Making a user complicit in the construction of an unfolding experience is a powerful technique for establishing engagement and commitment" [34, p. 95]. The speculation for the final interface was that when a child experiences agency in controlling the RDE, they are more likely to to be engaged and remain active in a conversation with it.

Based on observation and feedback, we found that that the interface was successful in engaging children in ongoing conversation across several domains: with the RDE (embodied); with their parent (social); and in post-study interviews (reflective). Rather than using a movable toy-like object, as was the case in the first and second interface iterations, the goal for the third design was to overcome previous issues with participants' mapping perception by using a scale model of the RDE structure as a tangible interface (Fig. 2). Across the surface of the model, coloured buttons afforded an uncomplicated capacity to generate lighting and sound based on the relative mapped position and colour of the button.

The clarity of fixed position and immediacy of button feedback successfully established awareness of control for the participants. The language of the interaction was unambiguous and each child was observed to understand their ability in controlling the responsiveness of the space. This immediate awareness provided a foundation for increasing complexity of the programmed response of the RDE: the longer a child pressed buttons in the same area of the interface, the more closely mapped the lighting and audio response was, relative to position on the surface of the RDE. In an attempt to establish conversation, as the system judges an awareness of the control based on repetition, mapping feedback is scaffolded through increasing complexity, requiring further attention from the participant.

Ultimately, if a child continues to interact with only a small section of the interface, control is removed and positional mapping no longer has a relationship to the buttons being pressed. The method of 'sabotage' allows the researcher to literally break repetitive interaction cycles and to produce an unexpected event. As sabotage occurs, several studies recorded instances of the child initiating communication with their parent, either for assistance ("can you try this?") or feedback ("what do you think will work?"); Thus, sabotage provides an interesting tool to disrupt a child's focus on the interface and to prompt a conversation across the social domain.

It is the position taken by this research that communication initiated from child to parent is a reflection of a sensory experience, motivated by exploring how interaction with the world takes place. Simon Penny sees this understanding of digital and social interaction as being linked: "We are first and foremost, embodied beings whose sensori-motor acuities have formed around interactions with humans, other living and non-living entities, materiality and gravity. We understand digital environments on the basis of extrapolations upon such bodily experience-based prediction" [34, p. 78]. When expectations of how a conversation should unfold is confounded, we experience what Susanne Bødker refers to in HCI research as a *breakdown* [10, p. 150]; an event that is unexpected for the participant and needs to be analysed further. This is not only useful to the researcher, but also the child as a unique centre of knowledge—compelled to reassess their agency, this becomes an activity through which social communication between child and parent often emerges.

3.3 Neurodiverse Conversation

The term *conversation* has been used in this chapter to describe different interactions across several domains. Despite this, the concept must be used carefully must be separated from its linguistic undertones. Influenced by the notion of neurodiversity, the usage of the term is deliberately accessible in that it doesn't require use of highly codified symbols, much like a language-based conversation would. *In My Language*, the widely viewed YouTube video created by Baggs [4], proposes an alternative perspective on how communication might look, sound or feel. In this video, Baggs suggests that rather than "being purposeless", her stereotypically autistic actions are in fact her "native language", as she sensorily engages with her environment [4]. This claim sits well within an embodied phenomenological way of experiencing the world [7], however, it remains opaque to the neurotypical view toward sensory stimulus and modalities of interaction and expression.

There is now a growing consent that defining autism through the use of statistical averages is misdirected. In support of this view, cognitive scientist, Jon Brock notes the importance of diversity in a recent article stating that: "heterogeneity is widely acknowledged by researchers. And yet autism research still focuses on the label itself, considering whether on average, people with autism are different from people who don't have that diagnostic label... autism isn't the average of people with an autism diagnosis" [8]. This research thus highlights uniqueness; the richness of experience which is important to consider in all research - not only through the contrast of those diagnosed with autism and so called *neurotypicals*, but in all human expression.

The term *neurodiversity* was coined by sociologist Judy Singer in the late 1990s [11, p. 2296]. Singer herself identifies as having Asperger syndrome and the neurodiversity movement, now embraced by self advocacy groups, uses the term to highlight that there is no such thing as autism in the singular, now encompassing many other cognitive conditions [11, p. 2296]. One of the goals of this research is to critically highlight what it means to communicate by acknowledging sensory interaction as a form of conversation and explore different modes of embodied expression.

4 Conversational Probes

The RDE can be considered as a critical design artefact: it is deployed to explore a rich variety of emergent responses from the user; a probe "to find out about the unknown—to hopefully return with useful or interesting data" [22, p. 18].

The concept of probes has been widely used in design and William Gaver in particular is well known for his development of *cultural probes* as a design method [19, 20, 22] which "tend to involve a single activity at a particular time and are not necessarily technologies themselves" [22, p. 18].

In terms of their use in a design process, cultural probes are specific in the activity that they explore and are not necessarily developed as an object for real-world use [22]. Hutchinson et al. reshape some of Gaver's ideas into what they term *technology probes*. Highlighting the multi-disciplinary approach that is integral to interaction design and HCI, technology probes aim to understand "the needs and desires of users in a real-world setting, the engineering goal of field-testing the technology, and the design goal of inspiring users and researchers to think about new technologies" [23, p. 17]. Although technology probes are open to emergent feedback, the approach is focused on the real-world use of a design artefact and therefore treats the participant as a user for testing, rather than being an ethnographic or narrative study as cultural probes can be interpreted.

This chapter presents an alternative approach to technology probes, which is particularly useful when working with at-risk populations, or studies in which assistive technologies are the focus. It is important that these groups are given an opportunity to express themselves wherever possible and as such, the *conversational probes* developed in this research not only pose questions about use, but like Pask's conversation theory, afford an opportunity to identify points of *agreement* or *understanding* between systems. Conversational probes are the first step in creating a common language of use, and have several distinguishing features:

- *Grounded*: As a question about embodied use directed toward the participant, they must be specific in their language. That is, they need to have a grounded understanding of the study population, so that the participant is able to communicate on their own terms.
- *Embodied*: Technology is not prescribed and is only used to support an examination of experience. Conversational probes are designed to improve our understanding of how a population embodies an experience in a situation, which occurs through the identification of events leading to *points of discussion* for reflection during and/or after interaction.
- *Accessible*: The term *conversational* places accessibility at the centre of design decisions. Participants are unique and important centres of knowledge, and as such, the probe artefact should augment the mode of communication most relevant to that group.

Conversational probes are not a method of analysis, though analysis of their use must identify and leverage any points of discussion elicited through their use. Following the experience-centred design approach of Wright and McCarthy, conversational probes put "the focus clearly on processes between people. It sees communication, knowledge, and identity as constructed in relationships between people, not within individuals" [43, p. 68]. Promoting an open approach to participants interacting with each other and the researcher(s), conversational probes are defined by the empathic design of a responsive artefact, which "can keep the diversity of voices alive" [44, p. 30]. The deliberately simple and play-like style of interaction in the RDE fostered immediate interaction leading to dialogues with the system. The resulting conversation with parents during the study and reflective discussion during post-study interviews suggest that this conversational focus for design can start to map experiential narratives in a rich and personal way, avoiding the exclusion and generalisation of marginalised groups, and "the implicit notion of 'user' in the singular" [11, p. 2296]. These approaches allowed for the participation to be analysed in the context of a broader sociocultural history of participants, rather than capturing use data without acknowledging the situational background of experience.

Conversational probes are a participatory design method, which recognises that "an understanding of interpretive and speculative approaches" is important to fully explore the rich and emergent experiences of participants. However, there is some danger that this approach can become *wicked*: when any measure of success may be different, depending on the perspective taken by any of the stakeholders; researchers, participants, or others that may pin their subjective hopes for the outcome(s) of the project [17, p. 2377]. In the case of the RDE study, the problem becomes manifold: the children, being the focus of the study, may find it difficult to isolate or communicate specific events for analysis, whether that is because of a difficulty with language or expressing abstract concepts. For that reason, this project incorporates the notion of *sabotage* as a method of motivating social communication in children.

4.1 Sabotage and the Breakdown

Sabotage in the RDE study was used to describe the act of withholding control from the participant, once it has been established (understood). In the studies described, there are two ways in which this may occur: if a participant repetitively triggers the same button(s), the system will become *bored* (inspired by Pask's *Musicolour*) and positional mapping will no longer function. Sabotage may also be triggered externally by the researcher, who is observing the study via a live video feed. The latter form of sabotage is activated to motivate child behaviour in response to the withholding of control.

Understandably, the term *sabotage* is a loaded one. Its use can be found most commonly in occupational therapy, where it is still contentious Mize [29]. The term is used here because it is effective in describing an act undertaken by agents external to the participant, whether they be the researcher or the system, and as a cybernetic framework, these systems are acknowledged and observed. Its purpose is in bringing focus to unconscious actions, for "[m]uch of the way we make sense of the world is what we might think of as tacit, part of the taken-for-granted of who we are" [43, p. 80]. There are two other concepts that describe a similar phenomenon which should be briefly outlined here: *expectation-violations* are used by Alcorn et al. [2] and Susanne Bødker's description of the *breakdown* in activity theory (2008, 149).

Expectation-violations have been observed by Alcorn in reference to the ECHOES project, which looks at autistic children's response to a screen-based virtual environment [17]. Alcorn sees these serendipitous moments (unplanned by the ECHOES project) as having the side-effect of initiating communication between child and adult, who share the interaction space. That these events were unplanned from the original project highlights an interesting problem when designing for autistic children, which Alcorn describes as "a valuable window into the interests and attentional focus of young children with [autism], illuminating the often significant gaps between the adult designer's intentions and the child's experience of the interaction" [2, p. 228]. Alcorn's sensitive observation adds further fuel to the premise that researchers need to be open to unique or emergent modes of interaction and communication.

Similarly, sabotage is used to trigger or encourage emergent behaviours [18]. By taking away scaffolded elements from a task or activity, the child may be compelled to ask for assistance, or find different ways of solving a problem. Naturally, care must be taken that this doesn't also lead to distress in a child with a developmental issue, or one who is generally anxious. Alcorn overcomes this issue by describing several design recommendations, one of which is to ensure that any discrepancy in the interaction is resolvable [2, p. 227].

Bødker's exploration of the breakdown as part of analysing activities is focused on the *use* of the artefact: "Breakdowns related to the use process occur when work is interrupted by something; perhaps the tool behaves differently than was anticipated, thus causing the triggering of inappropriate operations or not triggering any at all. In these situations the tool as such, or part of it, becomes the object of our actions" (2008, 150). The result of the breakdown is that the participant will reassess their ability to control or interact with the artefact—in this case the responsiveness of the RDE. In the majority of observational studies conducted in this project, the breakdown correlated with an initiation of social communication between child and parent.

Sinha et al. believe that for someone with autism, their environment can be an overwhelming place, "With compromised prediction skills, an individual with autism inhabits a seemingly "magical" world wherein events occur unexpectedly and without cause" [39, p. 1]. We believe that therefore the controlled interaction and precise environmental response of the RDE renders the conversational probe in this research a safe space for sabotage to occur.

Where the RDE differs from the ECHOES project is in the importance it places on physical immersion in an interactive experience. Alcorn's work suggests that novelty and surprise is useful in initiating communication, however ECHOES only looks at a screen-based interaction as a mode of interaction and therefore is limited in its ability to engage with sensory or alternative modes of communication. In ECHOES, the communication initiated by the child is required to be in the language of the adult; something that the RDE attempts to mitigate through the ability to engage in sensory communication initiated by the child.

4.2 A Participatory Process

It is important to remain mindful of the stakeholders that are involved throughout a research project. Particularly as the RDE study looks at an at-risk population, it is understandable that those who care for an autistic child (parents, teachers, therapists, et al.) would have hopes for the project that differ from the design researcher. By including these voices through reflective interviews (post-study reflection with participants; open discussion with therapists and practitioners), as well as using existing transdisciplinary research as a platform for dialogue, this project has striven to be a positive experience for all stakeholders.

Allison Druin submits that when designing new technologies for children, they should be consulted as part of that process [13]. As a take on cooperative inquiry, Druin's proposed framework not only recognises children as research partners, but maps out their "patterns of activity, communication, artifacts, and cultural relationships" [13, p. 593], establishing a rich picture of the interests of the child. This research benefits from acknowledging the power relationship between researcher, the child participant, and parent, and an appreciation that unique modes of communication have an important role to play in the design process. Particularly when working with an autism population who are likely to have very specific and pronounced responses to sensory stimulus [23], the interconnectedness of sensory experiences cannot be "rationally reduced to produce a single solution" [17, p. 2377] and each participant should be allowed to find meaning in the modality in which they choose to engage and interact.

5 Conclusion

The iterative design study presented in this chapter serves as a case study for taking a participatory approach to design through the use of conversational probes. Structured to uncover agreement and understanding between systems (either biological or digital), the method is based on features of a grounded understanding of the study population, and an embodied approach to experience and accessibility for all stakeholders. In the RDE project described in this chapter, this method proved to be particularly useful in opening dialogue between the researcher and participants, who were observed to engage with a responsive environment designed to elicit social interactions that may not result in linguistic expression. This case study suggests that where there is perceived power inequality between researcher and participant or when working with at-risk populations, conversational probes may assist in opening a study to conversation, through facilitation and awareness of different modes of interactive experience and embodied sensory communication.

References

1. Ackermann E (2001) Piaget's Constructivism, Papert"S constructionism: what"s the difference. Future of Learning Group Publication
2. Alcorn AM, Pain H, Good J (2014) Motivating children's initiations with novelty and surprise: initial design recommendations for autism In: ACM, New York, NY, USA, pp 28–225. http://doi.org/10.1145/2593968.2610458
3. Ashby WR (1956) An introduction to cybernetics. Chapman & Hall, London
4. Baggs A (2007) In: My language, vol 8
5. Boehner K, DePaula R, Dourish P, Sengers P (2005) Affect: from information to interaction, pp 59–68
6. Botts BH, Hershfeldt PA, Christensen-Sandfort RJ (2008) Snoezelen(R): empirical review of product representation. Focus on autism and other developmental disabilities, vol 23(3). SAGE Publications, pp 47–138. http://doi.org/10.1177/1088357608318949
7. Bower M, Gallagher S (2013) Bodily affects as prenoetic elements in enactive perception. Phenomenology and mind, July, 31–108
8. Brock J (2014) Connections: the elusive essence of Autism. SFARI Org Jan 28. https://sfari.org/news-and-opinion/columnists/jon-brock/2014/connections-the-elusive-essence-of-autism
9. Brown S (2013) The aesthetics of negotiation: using interactive technology to facilitate aesthetic choice of children with sensory processing disorders. In: changing facts, changing minds, changing worlds, Perth, pp 80–99
10. Bødker Susanne (1996) Creating conditions for participation: conflicts and resources in systems development. Hum-Comput Int 11(3):215–236. http://doi.org/10.1207/s15327051hci1103_2
11. Dalton NS (2013) Neurodiversity & HCI. CHI EA "13: CHI "13 Extended abstracts on human factors in computing systems, April. New York, NY, USA: ACM Request Permissions, p 2295. http://doi.org/10.1145/2468356.2468752
12. Dourish P (2001) 'Being-in-the-world': embodied interaction. In: Where the action is: the foundations of embodied interation. The MIT Press, pp 99–126
13. Druin A (1999) Cooperative inquiry: developing new technologies for children with children, New York, NY, USA: ACM, pp 99–592. http://doi.org/10.1145/302979.303166
14. Druin A (2002) The role of children in the design of new technology. Behaviour & Information Technology 21(1):1–25. http://doi.org/10.1080/01449290110108659
15. Dubberly H, Pangaro P (2009) What is conversation, and how can we design for it?. Interactions, 1 July. http://doi.org/10.1145/1551986.1551991
16. Dubberly H, Pangaro P, Haque U (2009) What is interaction? are there different types?." In: Dubberly H (ed) Interactions, 1 Jan. http://doi.org/10.1145/1456202.1456220
17. Frauenberger C, Good J, Keay-Bright W, Pain H (2012) Interpreting input from children: a designerly approach, New York, NY, USA. ACM Press, pp 86–2377. http://doi.org/10.1145/2207676.2208399
18. Ganz JB, Sigafoos Jeff (2005) Self-monitoring: are young adults with MR and autism able to utilize cognitive strategies independently? Educ Train Dev Disabil 40:24–33
19. Gaver B (2002) Designing for homo ludens, still. I3 Magazine, June
20. Gaver B, Dunne T, Pacenti E (1999) Design: cultural probes. Interactions, vol 6(1). ACM, pp 21–29. http://doi.org/10.1145/291224.291235
21. Heylighen F, Joslyn C (2001) Cybernetics and second-order cybernetics. In: Meyers RA (ed) Encyclopedia of physical science & technology, 3rd ed. Academic Press, New York, pp 1–24
22. Hutchinson H, Mackay W, Westerlund B, Bederson BB, Druin A, Plaisant C, Beaudouin-Lafon M et al (2003) Technology probes: inspiring design for and with families. ACM Request Permissions, Ft. Lauderdale, FL, USA, pp 5:17–24. http://doi.org/10.1145/642611.642616

23. Iarocci G, McDonald J (2005) Sensory integration and the perceptual experience of persons with Autism. J Autism Dev Disord 36(1):77–90. http://doi.org/10.1007/s10803-005-0044-3
24. Lipsky D (2011) From anxiety to meltdown: how individuals on the autism spectrum deal with anxiety, experience meltdowns, manifest tantrums, and how you can intervene effectively. Jessica Kingsley Publishers, London.
25. Jensen JF (1998) Interactivity: tracking a new concept in media and communication studies. Nordicom Rev, 185–204
26. Keay-Bright W, Howarth I (2011) Is simplicity the key to engagement for children on the autism spectrum? Personal and ubiquitous computing, vol 16(2). Springer, pp 41–129. http://doi.org/10.1007/s00779-011-0381-5
27. Maglione MA, Gans D, Das L, Timbie J, Kasari C (2012) Nonmedical interventions for children with ASD: recommended guidelines and further research needs. Pediatrics 130 (Supplement):169–178. http://doi.org/10.1542/peds.2012-0900O
28. McKee SA, Harris Grant T, Rice Marnie E, Silk Larry (2007) Effects of a snoezelen room on the behavior of three autistic clients. Res Dev Disabil 28(3):304–316. http://doi.org/10.1016/j.ridd.2006.04.001
29. Mize L (2008) A little frustration can go a long way …..using sabotage and withholding effectively to entice your toddler to talk. Teachmetotalk.com. 21 Apr. http://teachmetotalk.com/2008/04/20/a-little-frustration-can-go-a-long-way-using-sabotage-and-withholding-effectively-to-entice-your-toddler-to-talk/
30. Neely L, Rispoli M, Camargo S, Davis H, Boles M (2013) The effect of instructional use of an iPad® on challenging behavior and academic engagement for two students with Autism. Research in Autism spectrum disorders, vol 7(4). Elsevier Ltd, pp 16–509. http://doi.org/10.1016/j.rasd.2012.12.004
31. Pangaro P (2009) Cybernetics, conversation and applications for design: designing for conversation. Cambridge, pp 1–66
32. Parés N, Soler M, Sanjurjo A, Carreras A, Durany J, Ferrer J, Freixa P et al (2005) Promotion of creative activity in children with severe Autism through visuals in an interactive multisensory environment. ACM Press, New York, NY, USA, pp 16–110. http://doi.org/10.1145/1109540.1109555
33. Pask G (1968) A comment, a case history and a plan. In: Reichardt J Cybernetics, art and ideas, pp 76–99
34. Penny S (2011) Towards a performative aesthetics of interactivity. Fibreculture J 19 (December): 72–109
35. Peterson CC, Slaughter V, Peterson J, Premack D (2013) Children with Autism can track others' beliefs in a competitive game. Dev Sci 16(3):443–450. http://doi.org/10.1111/desc.12040
36. Pickering A (2010) Ontological theater. The cybernetic brain: sketches of another future, pp 17–36
37. Preece J, Rogers Y, Sharp H (2015). Interaction design: beyond human-computer interaction, 4 edn. Wiley
38. Rokeby D (1996) Transforming mirrors: subjectivity and control in interactive media. Davidrokeby.com. http://www.davidrokeby.com/mirrors.html
39. Sinha P, Kjelgaard MM, Gandhi TK, Tsourides K, Cardinaux AL, Pantazis D, Diamond SP, Held RM (2014) Autism as a disorder of prediction. In: Proceedings of the national academy of sciences, Oct, 1–6. http://doi.org/10.1073/pnas.1416797111
40. Stern N (2013) Interactive art and embodiment: the implicit body as performance. Gylphi Limited.
41. van der Aa O, Christine PDM, Pollmann MMH, Plaat A, van der RJ Gaag (2014) Computer-mediated communication in adults with high-functioning Autism spectrum conditions. arXiv.org
42. Wright P, McCarthy J (2008) Empathy and experience in HCI. ACM Press, Florence, pp 46–637. http://doi.org/10.1145/1357054.1357156

43. Wright Peter, McCarthy John (2010) Experience-centered design: designers, users, and communities in dialogue. Synth Lect Hum-Cent Inf 3(1):1–123. http://doi.org/10.2200/S00229ED1V01Y201003HCI009
44. Wright Peter, McCarthy John (2015) The politics and aesthetics of participatory HCI. Interactions. http://doi.org/10.1145/2828428
45. Yates GB (2002) A topological theory of Autism. Autismtheory.org. http://www.autismtheory.org/topotheory.html

FingerReader: A Finger-Worn Assistive Augmentation

**Roy Shilkrot, Jochen Huber, Roger Boldu, Pattie Maes
and Suranga Nanayakkara**

1 Introduction

Accessing printed reading material in an unstructured or unfamiliar environment is still a major challenge for people with visual impairments (VI). Whereas much of the printed material is not digitally accessible, many resort to using smartphone apps or simply asking for help from companions. Interviews with people with a VI [31] reveal that they struggle with focusing, aligning and even using reading assistive technology in settings such as in restaurants, on doctor appointments or reading mail items. In our experiments and interviews with persons with VI we validated these needs and problems and found a necessity for text-access technology that can overcome the hurdles of lighting, focus, aim and environment.

To this end we contributed the FingerReader, a finger-augmenting camera that looks at whatever the finger touches or points to. The major propositions of the FingerReader are: (i) using the finger for reading or pointing is well practiced within both sighted and non-sighted individuals, (ii) a finger-worn device creates a direct connection between the fingertip's strong tactile sensory capabilities and the directionality of the gesture, and finally (iii) camera and computer vision based algorithms can greatly benefit from the focused input as it is constrained to only what's underneath the finger or right in front of it.

The pointing gesture, flexing the index finger and pointing it at a thing, location or person, is a well practiced deictic gesture, rooted in the human gestural language and universally recognized across cultures and eras [19]. Pointing also carries many other symbolic meanings, such as signaling (e.g. in a classroom, or hailing a taxi),

R. Shilkrot (✉)
Computer Science Department, Stony Brook University, Stony Brook, NY, USA
e-mail: roys@cs.stonybrook.edu

J. Huber
Synaptics, Zug, Switzerland

R. Boldu · S. Nanayakkara
Singapore University of Technology and Design, Singapore, Singapore

P. Maes
Media Lab, Massachusetts Institute of Technology, Cambridge, MA, USA

© Springer Nature Singapore Pte Ltd. 2018
J. Huber et al. (eds.), *Assistive Augmentation*, Cognitive Science
and Technology, https://doi.org/10.1007/978-981-10-6404-3_9

or lexical meaning (e.g. the number one). The ubiquitousness of pointing makes it a strong candidate to augment technologically, since the entry barrier to performing the gesture is low, the social norms are lax, and it is well understood throughout society.

Using the pointing gesture for augmentation means capitalizing on the rich neural representation the index finger has in the somatosensory cortex of the brain. The tips of the index and middle fingers are the most highly dense areas of nerve endings, making them highly tuned to feeling an array of senses: tactile change in several frequencies, temperature, and pain [4]. The index finger is used by VI people to read braille, and by sighted people as a visual pointer for reading text in early learning stages. Utilizing this direct connection between the finger, fingertip and the brain, the FingerReader is a high sensory substitution modality between vision and tactility.

This article presents the comprehensive work performed on the FingerReader over the last 4 years towards its vision realization, beginning with the original FingerReader, on through the mobile version (MobiReader), the music reading version (MusicReader) and finally the latest development thrust to productization. We present the driving motivations, academic positioning and prior art, algorithmic components, design rationale and its implementation, as well as numerous user studies with visually impaired persons.

2   Related Work

Researchers in both academia and industry exhibited a keen interest in aiding people with VI to read printed text. The earliest evidence we found for a specialized assistive text-reading device for the blind is the Optophone, dating back to 1914 [6]. However the Optacon [16], a steerable miniature camera that controls a tactile display, is a more widely known device from the mid 20th century. Table 1 presents more contemporary methods of text-reading for the VI based on key features: adaptation for non-perfect imaging, type of text, User Interface (UI) suitable for VI and the evaluation method. Thereafter we discuss related work in three categories: wearable devices, handheld devices and readily available products.

Wearable devices. In a wearable form-factor, it is possible to use the body as a directing and focusing mechanism, relying on proprioception or the sense of touch, which are of utmost importance for people with VI. Yi and Tian [40] placed a camera on shade-glasses to recognize and synthesize text written on objects in front of them, and Hanif and Prevost's [10] did the same while adding a handheld device for tactile cues. Mattar et al. are using a head-worn camera [18], while Ezaki et al. developed a shoulder-mountable camera paired with a PDA [9]. Differing from these systems, we proposed using the finger as a guide [20], and supporting sequential acquisition of text rather than reading text blocks [30]. This concept has inspired other researchers in the community [33].

Table 1 Recent efforts in academia of text-reading solutions for the VI

Publication	Year	Interface	Type of text	Adaptation
Ezaki et al. [9]	2004	PDA	Signage	
Mattar et al. [18]	2005	Head-worn	Signage	Color, clutter
Hanif and Prevost [10]	2007	Glasses, tactile	Signage	
SYPOLE [22]	2007	PDA	Products, book cover	Warping, lighting
Pazio et al. [21]	2007		Signage	Slanted text
Yi and Tian [40]	2012	Glasses	Signage, products	Coloring
Shen and Coughlan [28]	2012	PDA, tactile	Signage	
Kane et al. [14]	2013	Stationery	Printed page	Warping
Stearns et al. [33]	2014	Finger-worn	Printed page	Warping
Shilkrot et al. [30]	2014	Finger-worn	Printed page	Slanting, lighting

Handheld and mobile devices. Mancas-Thillou, Gaudissart, Peters and Ferreira's SYPOLE consisted of a camera phone/PDA to recognize banknotes, barcodes and labels on various objects [22], and Shen and Coughlan presented a smartphone based sign reader that incorporates tactile vibration cues to help keep the text-region aligned [28]. The VizWiz mobile assistive application takes a different approach by offloading the computation to humans, although it enables far more complex features than simply reading text, it lacks real time response [3].

Assistive mobile text reading products. Mobile phone devices are very prolific in the community of blind users for their availability, connectivity and assistive operation modes, therefore many applications were built on top of them, however numerous specialized portable devices are also available. See Table 2 for a list of assistive products for reading text.

2.1 Finger Worn Cameras

The area of finger worn camera devices for interaction, not necessarily as assistive technology, is rapidly growing into a research agenda of it's own, albeit without notable consumer products yet in availability. The enduring work of Stetten et al. on FingerSight [12], first reported in 2006, tries to create an assistive finger-worn device to detect visual edges. The work of Nanayakkara and Shilkrot et al. spans a number of projects (not all cited here for brevity) into wearable assistive cameras to read text and recognize objects [20, 31], also occasionally serving as smartphone peripherals. Stearns et al. recently developed HandSight [33], which is geared directly at reading text with the finger. Other related work include the work of Rissanen et al. [25], which developed a smartphone peripheral camera for natural interaction with objects, and Yang et al., which created a miniature finger worn device that reacts to surface texture [36].

Table 2 Assistive mobile products for reading printed text

Name	Reference	Type
kNFB kReader	http://www.knfbreader.com	App
Text detective	http://blindsight.com	App
Text grabber	http://www.abbyy.com/textgrabber	App
StandScan	http://standscan.com	Device
SayText	http://www.docscannerapp.com/saytext	App
ZoomReader	http://mobile.aisquared.com	App
Prizmo	http://www.creaceed.com/iprizmo	App
LookTel	http://www.looktel.com	App
vOICe for Android	http://www.seeingwithsound.com	App
EyePal ROL	http://www.abisee.com	Device
OrCam	http://www.orcam.com	Device
Intel reader	http://reader.intel.com	Device
VizWiz	http://www.vizwiz.org	App
BeMyEyes	http://www.bemyeyes.org	App
TapTapSee	http://www.taptapseeapp.com	App

Camera-augmented fingers as an approach to assistive technology was also conceptualized earlier by designers without a technical implementation. Hedberg thought of the Thimble, a device to allow reading print and also braille [11], Lee designed the Reading Finger that reads barcodes [15], and both Munscher [1, 7] thought to use the finger as a point-and-shoot camera, literally.

The most relevant works in academia were already mentioned in [31], such as Kane et al.'s AccessLens [14], Yi's body of work [39] and Shen and Coughlan [28]. However other work not involving computer vision, such as El-Glaly's finger-reading iPad [8] and Yarrington's skimming algorithm [37], demonstrate the need to create an equilibrium between visual and non-visual readers by importing aspects of visual reading to assistive technology for VI persons.

3 FingerReader: A Wearable Reading Assistant

FingerReader supports persons with VI in reading printed text by scanning with the finger and uttering the words as synthesized speech. The device features hardware and software, including video processing algorithms, and two output modalities: tactile and auditory channels.

The design of the FingerReader is a continuation of our work on finger wearable devices for seamless interaction, namely the EyeRing [20, 30]. Exploring early design concepts with VI users revealed the need to have a small, portable device that supports free movement, requires minimal setup and utilizes real-time, distinctive multimodal response. Design explorations of form and function suggested such

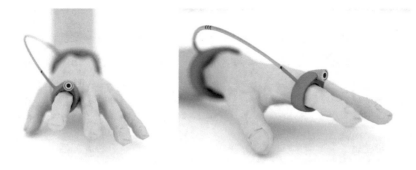

Fig. 1 Design vision. 3D modeling and rendering credits: Amit Zoran

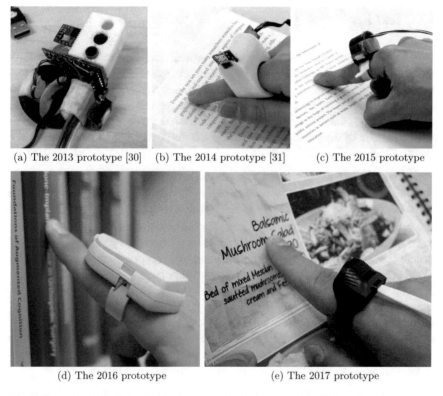

(a) The 2013 prototype [30]   (b) The 2014 prototype [31]   (c) The 2015 prototype

(d) The 2016 prototype   (e) The 2017 prototype

Fig. 2 Evolution of the FingerReader prototypes in the years 2013–2017

a device can be made aesthetically pleasing (Fig. 1), while keeping the camera in a fixed distance from the tip of the finger. The evolution of the laboratory prototypes (Fig. 2) led from a put-together mock up, to a fully enclosed device, decreasing in size and increasing comfort.

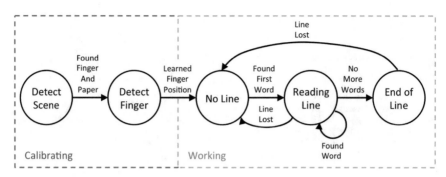

Fig. 3 Sequential text reading algorithm state machine

3.1 Text Reading Algorithm

The sequential text reading algorithm is comprised of a number of sub-algorithms concatenated in a state-machine (see Fig. 3), to accommodate for a continuous operation by a blind person. The first two states (*Detect Scene* and *Learn Finger*) are used for calibration for the higher level text extraction and tracking work states (*No Line, Line Found* and *End of Line*). Each state delivers timely audio cues to the users to inform them of the process. The algorithm in full detail can be reviewed in [31], therefore we only recount it in brief.

Scene and Finger Detection: The initial calibration step tries to ascertain whether the camera sees a finger on a contrasting paper. The input camera image is converted to the normalized-red channel: $nR = \frac{r}{r+g+b}$ that corresponds well with skin colors and ameliorates lighting effects. The image is matched to an example in a dataset of prerecorded typical images of fingers and papers. Once a stable match is achieved system deems the scene to be a well-placed finger on a paper. To detect the finger, an adaptive thresholding is performed and the top white pixel is considered a candidate fingertip point. During this process the user is instructed not to move, while our system collects samples of the fingertip location to build a location prior. The inlying fingertip detection guides a local horizontal *focus region*, located above the fingertip, within which the following states perform their operations. The focus region helps with efficiency in calculation and also reduces confusion for the line extraction algorithm with neighboring lines. See Fig. 4 for an illustration of this process.

Line Extraction: Within the focus region, we perform local adaptive binarization and selective contour extraction based on typical contour area sizes for characters. We pick the bottom point of each contour as the baseline point, and look for candidate lines by fitting line equations to triplets of baseline points, discarding extreme cases. We further prune by looking for supporting baseline points to the candidate lines based on distance from the line. We eliminate duplicate line candidates by binning and refine the equations based on their supporting points. We pick the highest scoring line as the detected text line. See Fig. 4 for an illustration.

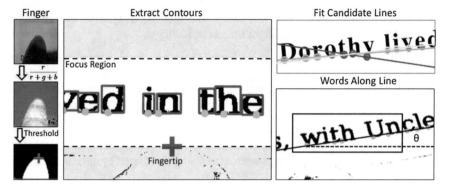

Fig. 4 Fingertip detection and text extraction

Word Extraction and Tracking: We employed the Tesseract OCR engine, set to only extract a single word and supply: the word, the bounding rectangle, and the detection confidence. Words with high confidence are retained, uttered out loud to the user, and further tracked using their bounding rectangle.

For tracking found words we use template matching over their image patches. Every successful match contributes to the bank of patches for that word. We constrain the template search region around the last position of the word while considering the predicted movement speed.

When the user veers from the scan line, detected using the line equation and the fingertip point, we trigger a gradually increasing auditory feedback. When the system cannot find more word blocks further along the scan line, it triggers an event and advances to the *End of Line* state.

3.2 Evaluation of the FingerReader

The central question that we sought to explore was how and whether FingerReader can provide effective access to print and reading support for VI users. Towards this end, we conducted a series of evaluations. First, we conducted a technical evaluation to assess whether the FingerReader is sufficiently accurate, and found that in perfect conditions the FingerReader's algorithm can recover over 93% of the words. In parallel, we performed an investigation of the usefulness of the different feedback cues with congenitally blind users. The results showed that participants preferred a tactile feedback compared to other cues (only audio or audio-and-tactile), and were able to recognized a gradual change in amplitude. One user reported that "*when [the audio] stops talking, you don't know if it's actually the correct spot because there's no continuous updates, so the vibration guides me much better.*"

We then used the results from these two fundamental investigations to conduct a qualitative evaluation of FingerReader's text access and reading support with 3 blind users. The methods, results and discussions thereof can again be read in depth in [31], and hereby we only highlight the major findings.

Qualitative User Study Findings We found that participants generally thought it was easy to access text with the FingerReader, however actual reading was considered less enjoyable and harder. Compared to other reading aids, participants were split between appreciating the immediacy of the FingerReader to the effectiveness of other aids.

- **Visual layout**: Restaurant menus and business cards were particularly challenging for the participants. Where some were specifically challenged by the multi-column layouts.
- **Audio feedback**: Some participants preferred an audio feedback to a tactile feedback, and mentioned that the choice of feedback could be better. One participant found it hard to navigate the different tones the FingerReader produced for line deviation and finger twisting/rotation.
- **Fatigue**: All participants reported that they would not use the FingerReader for longer reading sessions such as books, as it is too tiring. In this case, they would simply prefer an audio book or a scanned PDF that is read back, e.g. using ABBYY FineReader.
- **Serendipity**: Whenever any of the participants made the FingerReader read the very first correct word of the document, they smiled, laughed or showed other forms of excitement–every single time.
- **Efficiency over independence**: All participants mentioned that they want to read print fast and even "*when that means to ask their friends or a waiter around*". The FingerReader was marked with potential to help them towards independence, since they want to explore on their own rather than have others subjectively filter for them. However, efficiency in reading was consistently regarded as more important than independence.
- **Exploration impacts efficiency**: The former point underlines the potential of FingerReader-like devices for exploration of print, where efficiency is less of a requirement but getting access to it is. In other words, print exploration is only acceptable for documents where (1) efficiency does not matter, i.e. users have time to explore or (2) exploration leads to efficient text reading.
- **Layout navigation in an audio stream**: We found an indication that navigating text during the reading phase is comparable to the navigation in audio streams the device makes. The FingerReader recognizes words and reads them on a first-in, first-out principle at a fixed speed. Consequently, if the FingerReader detects a lot of words, it requires some time to read everything to the user. Stopping the finger movement to listen to the sound interrupted the interaction process and skewed the mental model of the blind user—the respective cognitive map of the document— specifically shaped through the text that is being read back.

On post-usage questionnaires, the overall experience with the FingerReader was rated as mediocre by all participants. They commented that this was mainly due to the synthesized voice being unpleasant and the steep learning curve.

4 MobiReader: A Mobile FingerReader

The second incarnation of the FingerReader was embodied in the MobiReader – a smartphone-based version of our software and a second iteration on the finger-worn camera design. Using a standard smartphone is key, since these are both prolific within the VI persons community and have ample computation power in recent generations. The MobiReader, designed as a peripheral device, could be made cheaper for using less components, and spare the user from purchasing a costly specialized device or even a new smartphone by simply adding external capabilities. Peripheral and smartphone-complementary devices are welcome in the VI community, a recent survey shows [38], as Bluetooth-coupled headsets and braille displays and keyboards are in wide use (Fig. 5).

To evaluate the MobiReader we designed a usability study with 10 VI persons in a lab setting, looking to estimate the potential success of the device as a mobile reading aid for printed material. Unlike former studies, here we contribute a quantitative assessment with a larger user base, and test the complete working system.

Our findings show that users were able to successfully extract an average of 74% of the words in a given piece of text when only provided with a feedback that told them how far away from the text line they were. The results demonstrate robustness in handling a range of standard font sizes, and that reading text within this range does not significantly hinder reading capability. The data also reveals insignificant

Fig. 5 The MobiReader camera peripheral

advantage for residual eyesight when using the MobiReader for reading, as some totally blind users actually had more success in reading than users with some residual vision.

The bulk of the details on the MobiReader, it's implementation and evaluation can be seen in [23, 29]. Hereby we describe the major differences the MobiReader made over the original FingerReader, as well as the results from further study into the proposed method of sequential text reading.

4.1 Improvements to the Device Hardware

Bearing resemblance to the FingerReader [31], the MobiReader is designed to be smaller and better adjustable to differently shaped fingers. The 3D-printed plastic case sports adjustable rubber straps and ergonomic design for adhering to the top of the finger. It also contains a considerably smaller camera module than that of the FingerReader, although not as small as the HandSight's NanEye [33]. The MobiReader, in contrast to FingerReader and HandSight, does not contain any vibration feedback capabilities and relies on audio cues alone, which allows it to be smaller and monolithic.

The camera module in use is analog; therefore a USB Video Class (UVC) video encoder is included with the system. The UVC interface allows the MobiReader to connect to practically any device with USB host capabilities and a modern operating system, smartphones included. This way the MobiReader could also be used as a peripheral by anyone carrying a smart device, e.g. a phone or an Android-enabled CCTV magnifier.

4.2 New Mobile User Interface on Android

We ported the implementation of the original computer vision algorithms for the Android platform in a new application (see Fig. 6), which also allows to control the hardware. Through the application, a variety of settings are available to the user that enables customizing the reading experience. Feedback settings can be adjusted: enabling and disabling candidate line feedback, distance, and angles, as well as customizing whether incoming words are read in their entirety or cut off when a new word is found. Speech rate, the speed at which words are read, can also be adjusted.

Android is an accessible operating system with built-in mechanisms to aid VI people to navigate the screens and interact with UI elements. With this new application the MobiReader is far easier for VI persons to adjust to their needs.

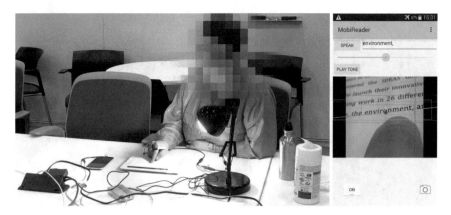

Fig. 6 *Left* P7 in midst reading with the MobiReader. *Right* Android app screen

4.3 Improvements to the Computer Vision Algorithm

The bulk of the algorithms used for the MobiReader are the ones used in [31], however our work contains a number of additional features and improvements. Existence of text (*No Text/Candidate Line* states) is determined by the number of qualifying character contours in the focus region, which is determined by the visible tip of the user's finger in the camera frame. If there are more than 2 qualifying characters that form a mutual baseline (tested by means of voting and fitting a line equation) the system transitions to *Candidate Line* state. In *Candidate Line* mode it will look for the first word on the candidate line via OCR.

The OCR engine, based on Tesseract [32], compensates for the distortion caused by the angle the finger takes with the paper. If the text is at an angle w.r.t the image, determined by the precomputed line equation, a 2D central rotation will correct it. Thereafter an intelligent trimming process will remove the whitespace surrounding the first word. We determine the first word by looking for large gaps in the x-axis projection of the words image patch (reducing the rectangular patch to a single row with the MAX operator on each column), similar to [33]. The trimmed patch is small enough to be quickly processed by Tesseract when set to the *Single Word* mode. OCR also does not occur on every frame, rather, only when new candidate words appear, greatly improving performance on our mobile processor.

The finger-tip detection algorithm of [31] was inefficient and expensive to execute in a mobile setting. We therefore introduced a coarse-to-fine method, where we start by analyzing an extremely downscaled (1% in number of pixels) version of the normalized-RGB image and later inspect the rough estimate in a small 100×60 pixel window to get a more precise reading. We also incorporate a standard Kalman filter to cope with noise in the measured fingertip point signal (see Fig. 11), which has a detrimental effect on the stability of the algorithms down the pipeline (Fig. 7).

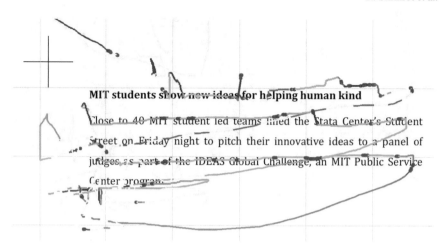

MIT students show new ideas for helping human kind

Close to 40 MIT student led teams lined the Stata Center's Student
Street on Friday night to pitch their innovative ideas to a panel of
judges as part of the IDEAS Global Challenge, an MIT Public Service
Center program.

Fig. 7 Text-Feedback aligned reading session. The *green-blue* marking are a visualization of the *Distance* feedback the user received, while the *red* marking is when word were spoken out loud. *Greener* hues correspond to a closer distance to the baseline, and *blue* hues farther. The *middle line* of the text was missed, however the user, a completely blind person, was still able to recover 73% of the words in total and spend over 51% of the time in line-tracking mode

4.4 Evaluation of the MobiReader

Work on the FingerReader did not include quantitative measurements and did not test an end-to-end system. The primary contribution of the MobiReader was a quantitative assessment of the complete system, including its computer vision subsystem, as used by a larger group of visually impaired persons. We recruited 10 participants to undertake monitored usage tasks and interviewed them about their experience. In total 10 reading tasks were designed to contain text of different sized fonts, layout and two variations of the audio feedback.

The feedback condition was the independent variable in a within-subjects design: Distance (D) and Distance + Angle (D + A). In 'Distance' the user hears a continuous feedback of how far their fingertip is from the line, and in 'Distance + Angle' the users hears 'Distance' and also a continuous feedback of the angle their finger makes with the line. Both feedbacks were given as sine waves of different pitches (Distance: 540–740 Hz, Angle: 940–1140 Hz). Each feedback condition was crossed with the tasks (5 tasks for D and 5 tasks for D+A) and fully counterbalanced to remove order bias.

4.5 Findings from the Quantitative Study

For metrics we designed three measurable effects: (i) Consecutive Score, which measures the amount of correctly and consecutively extracted words, (ii) Total Words

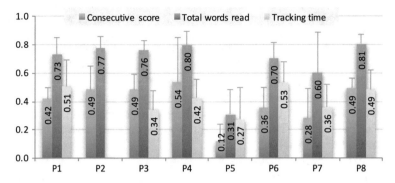

Fig. 8 Individual success in reading per participant

Read, which counts the number of correctly read words from the text without regard to sequence, as well as (iii) Tracking Time, which is simply the proportion of time the user spent in line-tracking mode versus line-searching mode.

On average our participants were able to correctly extract 68% (SD = 21%) of the words in the text, however some participants were able to extract up to 81% on average (see Fig. 8). The Distance feedback was somewhat better in helping users extract words from the text with 74% (SD = 18%) of the words on average, relative to 63% (SD = 22%) for Distance + Angle. Bigger font size only had a small positive effect w.r.t percent of extracting words (e.g. 72% for 11pt and 68% for 9 pt), but made a bigger impact in terms of the *Consecutive Score* (with 0.51 for 11 pt and 0.37 for 9 pt), which suggests, as one would expect, that larger font is easier to track.

As the *Consecutive Score* is not an absolute measurement, but rather a suggested model of the proficiency of a user in utilizing the MobiReader, it only can serve as a comparative measurement. As such, it does flush out the variance in users capabilities when it comes to feedback. Users not only extracted more words with only 'Distance' feedback turned on, they were also capable of extracting more consecutive words, with a score of 0.47 versus just 0.33 for 'Distance + Angle'.

The *Tracking time* measure provided little information as to how successful users were in reading, in spite of a correlation coefficient of 0.677 with *Total Words* and 0.526 with *Consecutive*. Interesting to note P6, the best participant in terms of time spent in the tracking modes (53% of the time), who was an Optacon user and understood very well the concept of text line tracking.

4.6 Qualitative Feedback

Open ended interviews with our participants revealed that all, save for one, did not appreciate the Angle feedback and were confused by multiplexing Distance + Angle (N = 8). Most users (N = 5) also mentioned the usage of the device causes excessive arm strain in keeping the finger and wrist straight and tense, as well as having to be very accurate and make very slight constrained movements (N = 3). Three users stated they would not use MobiReader to read long pieces of text, even though it was generally agreed that the device design was comfortable and small (N = 5). Some complained the overall reading process was slow (N = 3).

The prevailing reported strategy (N = 5) was to go top-to-bottom, i.e. finding the top line from the top of the page and working down to the next lines, as well as tracing backwards to the left to find the first word on the line; however backtracking was contested by some (N = 3).

Some users expressed dislike for the feedback in general, claiming the tone and increasing volume when straying from the line induced more panic than suggestion. At times this was reflected by large movements that could throw the user off the current line.

The evidence gathered in the MobiReader study largely corroborates the findings of the FingerReader studies. The auditory feedback was at times confusing, and users did far better with a simpler feedback, and fatigue was marked as a prevailing issue. While some users were very pleased with their newfound capability to explore text with their finger using a standard mobile device, there was still a general agreement that long pieces of text are better off read with an app or a dedicated device. We concluded that better algorithms for image analysis and text-tracking can alleviate some of the problems and create a smoother experience.

5 MusicReader: A Printed Music Sheet FingerReader

The latest evolution of the FingerReader's sequential reading algorithms is the MusicReader: a printed sheet music reading algorithm. While reading sheet music shares many traits of reading plain printed text, it also brings about many challenges to solve. Our work resulted in a unique Optical music recognition (OMR) algorithm that is able to sequentially trace and read back the note it encounters to a reader with VI. The motivation for our work came from interviews we held with musicians with VI, which revealed that in order to read printed music sheets they rely on human transcribers or scanning using specialized stationary equipment that often produces recognition errors. Printed music sheets for musicians with VI in accessible formats such as music braille, is generally considered expensive, rare and inefficient for reasons of portability. For musicians with VI participating in music classes or band sessions, this issue creates a barrier between them and their sighted colleagues, as they are not as independent. The MusicReader strives to enable music readers with VI to access non-instrumented paper musical notation sheets in a mobile context, and level the playing field with their sighted peers. Full details of the MusicReader's implementation and evaluations can be seen in [29], and hereby we highlight the main differences over the MobiReader and share only the key findings from the study with musicians with VI (Fig. 9).

Fig. 9 The MusicReader. *Left* The finger-wearable camera design. *Right* A blind musician in a reading session

Needs Of VI Music Readers. Musicians with VI looking to learn new music mostly use braille music, learning by ears, or digitized music sheets in accessible digital formats. Braille music is a relatively prolific accessible format for encoding musical information based on the braille character set, however it presents a number of acute challenges. Braille music is expensive to produce and thus also to purchase, since it is a niche format for musical notation, which leads to a small offering of music translated to braille notation [13]. In addition, there are only few qualified braille music transcribers who can produce braille music,[1] and few teaching institutions for braille music exist. Braille music translation also results in very heavy and large "printed" books, which is another usability factor impeding accessibility. Although some VI musicians have outstanding aural skills that help them to learn new materials quickly by ear, it's not the case for every musician with VI, and there is information, such as finger markings, that entirely cannot be retrieved simply from listening.

These challenges with learn-by-ear and braille music lead some musicians with VI to learn new music by digitizing printed music sheets using OMR, however this too is not free of limitations. Operating a flatbed scanner requires experience using a screen-reader and the specific printed music digitization software (e.g. SharpEye[2] or SmartScore[3]). The physical setup for scanning is also important for a successful sightless operation, therefore it is usually done in a recognizable location, such as a specialized room or at home. But even in perfect scanning conditions, a properly scanned page will often result in errors in the OMR process. Furthermore, users with VI would not be able to recognize and correct these errors independently since they cannot refer to the original printed music sheet. Scanning in a different scenario, such as using a mobile phone, presents problems of aim, focus and alignment, but more importantly—such mobile music scanning applications for the VI are hardly in existence.

---

[1]The Library of Congress lists 70 braille music transcribers US-wide: http://www.loc.gov/nls/music/circular4.html.

[2]http://www.visiv.co.uk/.

[3]http://www.musitek.com/.

Problems with existing solutions are more acute in social situations such as a classroom or a band, where musicians with VI are expected to access music handouts in the same way their sighted peers do. Accessible workbooks and pre-digitized music do exist, however readers with VI are confined to this content and cannot spontaneously access printed material. Consequently, a music reading solution tailored for the VI to use in a mobile context could provide them with a way to better integrate into the learning and communal playing environment.

Existing Mobile OMR Solutions. The MusicReader is similar to Gocen [2], a music reading system where the user is allowed to scan a stave notation line using a handheld camera and hear the notes played in real time. However Gocen does not recognize any symbols other than full stemless notes, has no memory of notes outside of its immediate view, as well as it doesn't provide non-visual feedback on the scanning other than playing the note. Another related work is onNote by Yamamoto et al. [35], which uses the index finger as an access pointer to different parts of a paper-printed music sheet and changing the nature of playback with visual projected feedback. The sheets in onNote are scanned into the system beforehand and only matching to the existing database of pre-processed sheets can be performed. The advent of computationally capable smartphones enabled performing OMR on the phone itself within apps [17, 34], however these are not geared towards VI people and provide only visual feedback.

5.1 User Interface and Feedback

Similar to the feedback the MobiReader and FingerReader provide, the MusicReader has two main feedback components: scanning feedback via audio tones, and content (music notation) feedback via speech.

- **Speech Feedback**. Each note encountered in the scan is translated to duration and pitch [5]: "eighth", "quarter", or "half", followed by the pitch class in latin letters (CDEFGAB), and finally the octave ("3", "4" or "5"), for example "eighth-D4". Accidentals are uttered as class and pitch (e.g. "Flat-D4"), and symbols without pitch simply utters the word (e.g. "bar", "quarter rest").
- **Tonal Feedback**: The tonal feedback guides the user in scanning a line of stave notation music. The goal is to help the user keep the finger-camera pointing at roughly the middle of a line, via feedback that describes the distance from the center of the line. The major difference from before is that the feedback is binary: above the line (a high C note), and below the line (a low C note), instead of a continuous varying tone to describe the distance. This simplification greatly reduces cognitive load, as there is no tonal feedback when roughly centered so users can concentrate on the notes utterances. When the system cannot detect any line in the image it emits a quieter G note tone.

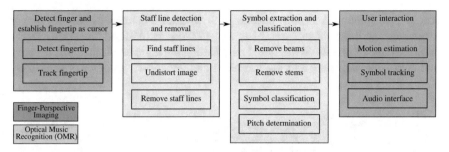

Fig. 10 The processing and interaction pipeline in our assistive music reading system

5.2 Computer Vision Algorithm for Finger-Perspective OMR

The MusicReader's computer vision system considers a unique approach for extracting musical information from printed sheets. A local view from the finger perspective is generally considered easier to computationally analyze, but it also introduces additional problems that do not exist in other OMR systems: using the finger as the cursor for the analysis, handling a moving view of the page, and providing feedback on the scanning operation itself rather than the content alone (the musical symbols). These issues augment the traditional requirements from an OMR pipeline, which are also included in our system: staff line detection and removal, segmentation, classification and more (see Fig. 10). With respect to Optical Character Recognition (OCR), OMR is considered harder as the symbols are often converged and multiplexed rather than clearly demarcated as text characters.

Staff Lines Detection, Removal and Tracking. The fingertip location in the image, calculated as was done in the MobiReader (see Fig. 11), allows us to process just a small region of interest where we look for the staff lines using [26]. Assuming the staff lines run from the left to the right extremities of the small region we pick the left-to-right lines that have the most black pixels along them, and then perform binning based on the line's intercept to finally converge to the 5 unique staff lines. To validate, we calculate the distance between neighboring lines as well as their angles, where a good line detection is when all measurements agree within a small variation. For further calculations we extract the staff line space (SLS) and staff line height (SLH) from the detected staff lines. In subsequent frames, we search for new staff lines only within a small region around the previously found lines, to speed up computation.

The staff lines impose a near-uniform 2D rotation of the image, although in some cases, depending on the perspective distortion, the lines disagree on the angle. Using the inverse 2D rotation roughly rectifies the symbols for proper classification (see Figs. 12c, and 13d).

Staff Line Removal. For removing the staff lines and keeping the musical symbols intact we use a Hidden Markov Model (HMM) (see Fig. 12). We sequentially scan the staff line to determine at each point whether it belongs to the staff line or a symbol. The hidden states we utilize are: {STAFF, SYMBOL, SYMBOL THIN, NOTHING}. Observations are based on counting the number of black pixels above and below the staff line, and the transition and emission matrices were manually

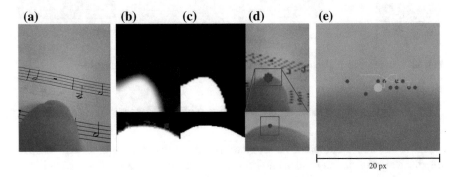

Fig. 11 Fingertip detection process. **a** Original image, Coarse-to-fine detection of the fingertip: **b** detected skin probably map, **c** binarized using Otsu method, **d** detected fingertip point, e Kalman filtering of the measurement point (*green*—filtered result, *red*—measurements with noise)

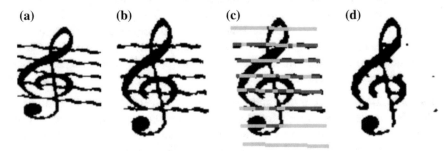

Fig. 12 Classification of staff lines for removal. **a** the original binarized image, **b** the undistorted view, **c** annotated staff lines from the HMM, and **d** the staff lines removed with minimal damage the symbol

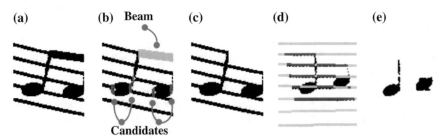

Fig. 13 **a** the binarized input region, **b** detecting consecutive segments of viable height, **c** removing the segments classified as beams, **d** staff line removal on the rectified image, **e** result with beam and staff lines removed, ready for note classification

trained from a number of annotated examples. To discover the annotation for a new staff line we calculate the observation sequence and running the Viterbi algorithm, which is then used to remove all pixels in the STAFF or NOTHING state according to the staff line height (SLH). We used [27] for the HMM implementation.

Beam and Stem Removal To correctly classify beamed groups of notes (e.g. connected eighth notes), we must remove the connecting beam. We detect vertical segments in the image that are likely to be part of a beam (based on their pixel length), and look for consecutive overlapping segments whom centers also converge on a line, since a beam is always a straight line. We finally remove the selected beam segments by painting over the pixels (see Fig. 13).

Many of the note symbols arrive at this point of the OMR pipeline with a stem (the vertical line going above or below the note head): half notes, quarter notes and also connected eighth notes after having their beam removed. However for a simple pitch and duration classification we keep only the note head, which we find from the analyzing the y-axis projection (reduce-sum operation on the rows).

Symbol and Pitch Classification. Once we obtain a clean symbol we use geometric features of the contour with a decision tree classifier to classify the symbol to its type (e.g. note head, accidental, bar line, etc.). Inspired by [24], we use the following features: width, height, area, ratio of black versus white pixels, and 7 Hu moments. The decision tree was trained with a dataset of 1170 manually classified note symbols from a training set of images.

To determine the pitch for note heads we consider the the center of mass and for incidentals (sharps and flats) we use the central point from bounding rectangle. If the symbol central point is within 15% of the SLS to one of the staff lines, we deem the symbol to lie on that line and assign an octave and pitch. See Fig. 14 for an illustration of this process.

5.3 Evaluation of the MusicReader with VI Musicians

To evaluate the performance and usefulness of the system we performed a controlled user study with VI musicians. The goal of the study was to assess the feasibility of the MusicReader to assist in reading a printed music sheet in an unstructured environment, simulating the real situation a person would wish to use the device. We recruited 5 participants from a pool of volunteer VI musicians, and an additional VI musician volunteered to act as a pilot user for the study.

Study participants used two printed music sheets in standard staff notation for reading. The melodies on the sheets were simple arrangements (no harmony) of the following standards: "Happy Birthday", "Greensleeves", "Over the Rainbow" and "Amazing Grace" (see Fig. 15). Participants had a chance to try and read the two sheets, and were questioned about whether they can recognize the melodies in them. The read notes and audio feedback events were recorded by the software on the PC along with timestamps, and were later analyzed as quantitative measures.

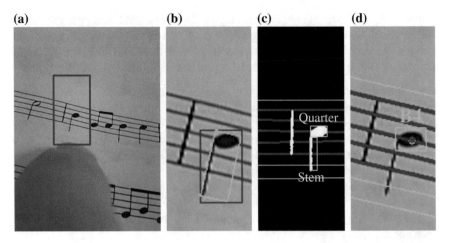

Fig. 14 Symbol pitch and type classification process: **a** original image with region of interest, **b** detected contour to classify, **c** the segmented contour of note-head and stem, **d** the classified note-head based on the position on the staff lines

Fig. 15 One of the sheets the participants used for reading, with arrangements of "Happy Birthday" and "Greensleeves"

5.4 Key Findings from the Quantitative and Qualitative Study

The MusicReader is a novel approach in the domain of mobile assistive technology for musicians with VI, where most prior work did not attempt to tackle non-visual reading of printed music. As such, study participants had mixed comments about its utility, although there was a positive consensus about its potential.

For the quantitative part, the results show most users were able to cover roughly %35 of the notes from the test sheets in the allotted time for reading independently (20 min). All users were stopped by the examiner at the end of the time frame, therefore given additional time they would continue to read more notes. Thus the result on the proportion of notes read is severely skewed.

In the results of the qualitative part, which consisted of questionnaires and open interviews, participants did not think the MusicReader was easier than other music reading aids, although most reported that they do not know of similar aids. All participants save for one stated they would require an expert to help them operate the device but felt there weren't many things to learn.

Learning-by-ear While most of the participants in our experiment noted the MusicReader is an intriguing technology that would be useful when fully developed, all participants agreed that learning-by-ear is still the best tool they have to access music.

> I would love to be able to read music, but I still consider having aural skills, the ability to learn a piece by ear and play it back, a very important tool. It could be used in combination with reading, and I still think being able to read is a good thing.

On the other hand, our interviewees reported of numerous situations where learning-by-ear is impossible or impractical: band practice, working with a conductor, in the classroom and while teaching. In these situations, according to our participants, musicians with VI at a disadvantage even in spite of their technical abilities.

> Not being able to access printed music material knocks blind people out of a big segment of the market. I don't think I could go audition for the BSO [Boston Symphony Orchestra], even though I think I have the chops to at least play 3rd or 4th trumpet for the BSO, but they want you to be able to sight-read. Not to be able to work on-the-fly like that is a really big problem. [...] At the moment I would not be able to teach beginners that don't know how to read, but I can certainly teach them how to play. I feel like that's something that keeps me from teaching beginners privately.

Finger Positioning and Aiming. Most study participants had problems of aiming the device and maintaining the right angle for proper reading. This problem in the MusicReader is key as musicians with VI read with the specific goal of playing their instrument, and therefore can spare, at most, one hand for reading depending on the instrument they play.

> P3: I can't have my left hand off of the trumpet, I must hold it. [...] I have to have both hands on the instrument

Study participants were also concerned with getting a very quick and precise reading, and did not have much patience towards learning the hand positioning or maintaining it for long. We conclude that both the imaging hardware as well as the software may need to improve to overcome this problem. The camera lens could be of a wider angle, and the algorithms to find the fingertip and staff lines could have a much higher tolerance towards skewed views (Fig. 16).

> P2: [...] the finger needs to be in a very specific position, there should be a better way. The angle was not directly straight with the paper, and I can't see the paper.

Tonal Signals for Reading Music Some participants reported of an increased cognitive load when listening to the assisting tones while trying to mentally reconstruct the music only from the spoken names of the notes. In reading music with the MusicReader this issue of mental interference may be more severe than in reading printed text, which suggests tonal feedback may be less effective. Participants suggested we incorporate tactile feedback to circumvent this issue.

Fig. 16 User study participants in midst reading a music sheet

P2: The notes are good for feedback, but if you're thinking about the music - that's confusing. Maybe it shouldn't be in the music range, not C,G and C if I am reading something in a C scale.

The findings from our study point both to the potential of the MusicReader as a mobile assistive technology and to the usability obstacles of such an approach. Reading printed music for musicians with VI is not a special case of reading printed text but rather a new problem class. In many situations music is read with the goal of immediately playing it, often in a group setting with other musicians, which requires a fast response, high accuracy and less than ideal reading conditions. Reading music also requires the reader to mentally reconstruct the music, which can interfere with any audio feedback from the system. Nevertheless, some elements of reading music are similar to reading text such as locating oneself in the page.

6 Latest Design Iterations of the FingerReader Hardware

Since the early lab prototypes of the FingerReader (Fig. 2), efforts were devoted to improve the usability of the FingerReader device. Further iterations on the camera electronics have reduced the size considerably, and went hand in hand with industrial design iterations on the casing and materials. Finally, the new device has a much smaller form factor and higher comfort (see Fig. 2e). The onboard miniature camera operates at VGA resolution (640 × 480 pixels) and 30 frames-per-second, and connects via standard USB to a PC, smartphone or smartwatch. The latest version of the FingerReader design was produced in order of thousands, and many devices were distributed to users and organizations pending a large-scale user study.

7 Conclusion

The FingerReader is a unique assistive augmentation interface for reading by point-ing. Over the last 4 years we led many design and development iterations, demonstra-tions and evaluations to assess the FingerReader's feasibility as an assistive technol-ogy for visually impaired persons. Results of numerous user studies with the target audience—persons with VI—show clear potential for FingerReader to be used in exploring printed documents. However, there is an obvious need to improve on a number of fronts: the computer vision algorithms must improve to allow for more intuitive usage with less guidance, the feedback mechanisms must match the appli-cation as well as support and not obstruct the content.

Acknowledgements We would like to acknowledge the people who were directly involved in the ideation, creation and evaluation of the FingerReader: Connie Liu, Sophia Wu, Marcelo Polanco, Michael Chang, Sabrine Iqbal, Amit Zoran, K.P. Yao. We would like to acknowledge the help of the VIBUG group in MIT for testing and improving the FingerReader.

References

1. Ubi-Camera (March 2012). http://www.gizmodo.in/gadgets/Finger-Camera-Lets-You-Frame-a-Shot-Like-a-Pompous-Director/articleshow/19139922.cms
2. Baba T, Kikukawa Y, Yoshiike T, Suzuki T, Shoji R, Kushiyama K, Aoki M (2012) Gocen: a handwritten notational interface for musical performance and learning music. In: SIGGRAPH 2012 emerging technologies. ACM
3. Bigham JP, Jayant C, Ji H, Little G, Miller A, Miller RC, Miller R, Tatarowicz A, White B, White S, Yeh T (2010) VizWiz: nearly real-time answers to visual questions. In: Proceedings of the 23nd annual ACM symposium on user interface software and technology, UIST '10, New York, NY, USA. ACM, pp 333–342
4. Byrne JH, Dafny N (1997) Neuroscience online: an electronic textbook for the neurosciences. The University of Texas Medical School at Houston, Department of Neurobiology and Anatomy
5. Crombie D, Dijkstra S, Schut E, Lindsay N (2002) Spoken music: enhancing access to music for the print disabled. In: Computers helping people with special needs. Lecture notes in computer science, vol 2398. Springer, Berlin, pp 667–674
6. d'Albe EEF (1914) On a type-reading optophone. Proc R Soc Lond A 90(619):373–375
7. David M (Oct 2007) Every stalker's dream: camera ring
8. El-Glaly YN, Quek F, Smith-Jackson TL, Dhillon G (2012) It is not a talking book;: it is more like really reading a book! In: Proceedings of the 14th international ACM SIGACCESS conference on computers and accessibility, ASSETS '12, New York, NY, USA. ACM, pp 277–278
9. Ezaki N, Bulacu M, Schomaker L (2004) Text detection from natural scene images: towards a system for visually impaired persons. Proc ICPR 2:683–686
10. Hanif SM, Prevost L (2007) Texture based text detection in natural scene images—a help to blind and visually impaired persons. In: CVHI
11. Hedberg E, Bennett Z (Dec 2010) Thimble—there's a thing for that
12. Horvath S, Galeotti J, Wu B, Klatzky R, Siegel M, Stetten G (2014) FingerSight: fingertip haptic sensing of the visual environment. IEEE J Trans Eng Health Med 2:1–9

13. Jacko VA, Choi JH, Carballo A, Charlson B, Moore JE (2015) A new synthesis of sound and tactile music code instruction in a pilot online braille music curriculum. J Vis Impair Blindness (Online) 109(2):153
14. Kane SK, Frey B, Wobbrock JO (2013) Access lens: a gesture-based screen reader for real-world documents. In: Proceedings of the SIGCHI conference on human factors in computing systems, CHI '13, New York, NY, USA. ACM, pp 347–350
15. Lee H (Sept 2011) Finger reader
16. Linvill JG, Bliss JC (1966) A direct translation reading aid for the blind. Proc IEEE 54(1):40–51
17. Luangnapa N, Silpavarangkura T, Nukoolkit C, Mongkolnam P (2012) Optical music recognition on android platform. In: Advances in information technology. Communications in computer and information science, vol 344. Springer, Berlin, pp 106–115
18. Mattar MA, Hanson AR, Learned-Miller EG (June 2005) Sign classification using local and meta-features. In: CVPR—workshops. IEEE, p 26
19. McNeill D (2000) Language and gesture, vol 2. Cambridge University Press
20. Nanayakkara S, Shilkrot R, Yeo KP, Maes P (2013) EyeRing: a finger-worn input device for seamless interactions with our surroundings. In: Proceedings of the 4th augmented human international conference, AH '13, New York, NY, USA. ACM, pp 13–20
21. Pazio M, Niedzwiecki M, Kowalik R, Lebiedz J (2007) Text detection system for the blind. In: 15th European signal processing conference EUSIPCO, pp 272–276
22. Peters J-P, Thillou C, Ferreira S (2004) Embedded reading device for blind people: a user-centered design. In: Procedings of the ISIT. IEEE, pp 217–222
23. Polanco MRII (2015) Mobireader: a wearable, assistive smartphone peripheral for reading text. Master's thesis, Massachusetts Institute of Technology
24. Rebelo A, Fujinaga I, Paszkiewicz F, Marcal ARS, Guedes C, Cardoso JS (2012) Optical music recognition: state-of-the-art and open issues. Int J Multimedia Inf Retrieval 1(3):173–190
25. Rissanen MJ, Fernando ONN, Iroshan H, Vu S, Pang N, Foo S (2013) Ubiquitous shortcuts: mnemonics by just taking photos. CHI '13 extended abstracts on human factors in computing systems, CHI EA '13. New York, NY, USA. ACM, pp 1641–1646
26. Roach JW, Tatem JE (1988) Using domain knowledge in low-level visual processing to interpret handwritten music: an experiment. Pattern Recognit 21(1):33–44
27. Sand A, Pedersen CNS, Mailund T, Brask AT (2010) HMMlib: a C++ library for general hidden Markov models exploiting modern CPUs. In: 2010 ninth international workshop on parallel and distributed methods. IEEE, pp 126–134
28. Shen H, Coughlan JM (2012) Towards a real-time system for finding and reading signs for visually impaired users. In: Computers helping people with special needs. Springer, pp 41–47
29. Shilkrot R (2015) Digital digits: designing assistive finger augmentation devices. PhD thesis, Massachusetts Institute of Technology
30. Shilkrot R, Huber J, Liu C, Maes P, Nanayakkara SC (2014) FingerReader: a wearable device to support text reading on the go. In: CHI EA. ACM, pp 2359–2364
31. Shilkrot R, Huber J, Meng Ee W, Maes P, Nanayakkara SC (2015) Fingerreader: a wearable device to explore printed text on the go. In: Proceedings of the 33rd annual ACM conference on human factors in computing systems, CHI '15, New York, NY, USA. ACM, pp 2363–2372
32. Smith R (2007) An overview of the tesseract OCR engine. In: ICDAR, pp 629–633
33. Stearns L, Du R, Oh U, Wang Y, Findlater L, Chellappa R, Froehlich JE (Sept 2014) The design and preliminary evaluation of a finger-mounted camera and feedback system to enable reading of printed text for the blind
34. Viktor L (Nov 2014) iSeeNotes—sheet music OCR!
35. Yamamoto Y, Uchiyama H, Kakehi Y (2011) onNote: playing printed music scores as a musical instrument. In: Proceedings of UIST. ACM, pp 413–422
36. Yang XD, Grossman T, Wigdor D, Fitzmaurice G (2012). Magic finger: always-available input through finger instrumentation. In: Proceedings of the 25th annual ACM symposium on user interface software and technology, UIST '12, New York, NY, USA. ACM, pp 147–156

37. Yarrington D, McCoy K (2008) Creating an automatic question answering text skimming system for non-visual readers. In: Proceedings of the 10th international ACM SIGACCESS conference on computers and accessibility, Assets '08, New York, NY, USA. ACM, pp 279–280
38. Ye H, Malu M, Oh U, Findlater L (2014) Current and future mobile and wearable device use by people with visual impairments. In: Proceedings of the SIGCHI conference on human factors in computing systems, CHI '14, New York, NY, USA. ACM, pp 3123–3132
39. Yi C (2010) Text locating in scene images for reading and navigation aids for visually impaired persons. In: Proceedings of the 12th international ACM SIGACCESS conference on computers and accessibility, ASSETS '10, New York, NY, USA. ACM, pp 325–326
40. Yi C, Tian Y (2012) Assistive text reading from complex background for blind persons. In: Camera-based document analysis and recognition. Springer, pp 15–28

Conclusion

Assistive Augmentation is a budding field of research and still in its infancies. Catering to humans who seek assistive or enhancing devices, it requires many disciplines to work together. The major goal of this edited volume is to illustrate-by-example the core areas of Assistive Augmentation as a research discipline and to establish a common ground for the growing community. The transdisciplinary nature of the field is underlined by case studies presented in this book. At the same time, these case studies also establish emerging challenges. Two core areas of Assistive Augmentation were mapped to the respective parts in this book: *Sensory Enhancement and Substitution*, as well as *Design for Assistive Augmentation*.

In the first part, the notion of "Augmented Sensors" was introduced, which describes modes of sensory remapping, enhancement and conceptual creation of new sensing modalities. Chapter 3 highlighted challenges of sensory remapping for deaf people, scaffolding music listening and music making practices. While Chap. 4 contributes insights into augmentation technologies for future industry workers as flexible production environments, Chap. 5 exemplifies the potential of Assistive Augmentation technology to enhance experiences (here: reading).

The second part of the book focused on designing for Assistive Augmentation, embracing physical environments, social spaces and enabling devices. Chapter 7 focused on design challenges for our ageing population and how the design of augmentation technologies, as well as domestic environments needs to cater to ageing in place. Chapter 8 presented a responsive sensory environment, pointing out design requirements for augmented social spaces; here targeting communication between autistic children and their parents. Chapter 9 depicted the design process of a concrete Assistive Augmentation technology over the course of half a decade: the FingerReader, a finger-worn device equipped with a camera, auditory and haptic feedback that assists people with visual impairments in accessing printed information.

The focus in this volume was on developing, designing, understanding and studying assistive augmentation technologies. Other important core areas of this young interdisciplinary field were not addressed, such as ethical issues, quality of augmentations and their appropriations, as well as amplifying and augmenting

© Springer Nature Singapore Pte Ltd. 2018
J. Huber et al. (eds.), *Assistive Augmentation*, Cognitive Science
and Technology, https://doi.org/10.1007/978-981-10-6404-3

human perception. The latter was a focus in a recent workshop held at the 2017 ACM CHI Conference on Human Factors in Computing Systems [1]. These ongoing discussions in academic communities underline the importance of investigating, shaping and defining the intersection of assistive technologies and human augmentations. Academic research is one avenue that must be pursued, with work being disseminated at dedicated conference series such as Augmented Human [2]. Other avenues that highlight and demonstrate the potential of Assistive Augmentation technology include for instance sports, as discussed within the Superhuman Sports Society [3]. Most recently, the Cybathlon was held for the very first time in 2016. Athletes with *"disabilities or physical weakness use advanced assistive devices [...] to compete against each other"* [4]. The 2016 edition featured six disciplines, ranging from brain-computer interface races to powered exoskeleton and wheelchair races, showcasing the breadth of augmentations that were allowed into the competition to assist athletes.

Drawing from these recent developments in academia and beyond, we believe the potential for Assistive Augmentation as a research field is enormous and the timeliness of the presented research projects and explorations is perfect. We are eagerly looking forward to future developments and to witness Assistive Augmentation pushing our physical, sensorial and cognitive abilities.

References

1. Albrecht Schmidt, Stefan Schneegass, Kai Kunze, Jun Rekimoto, and Woontack Woo, "Workshop on Amplification and Augmentation of Human Perception," in *Proceedings of the 2017 CHI Conference Extended Abstracts on Human Factors in Computing Systems*, 2017, pp. 668–673
2. "Augmented Human Conference Series." [Online]. Available: http://www.augmented-human. com/. [Accessed: 01-Sep-2017]
3. "Superhuman Sports Society." [Online]. Available: http://superhuman-sports.org/. [Accessed: 01-Sep-2017]
4. "Cybathlon," *Cybathlon - moving people and technology*. [Online]. Available: http://www. cybathlon.ethz.ch/. [Accessed: 01-Sep-2017]

Printed in the United States
By Bookmasters